ORBITAL MECHANICS

ORBITAL MECHANICS
Theory and Applications

TOM LOGSDON

A Wiley-Interscience Publication

JOHN WILEY & SONS, INC.

New York / Chichester / Weinheim / Brisbane / Singapore / Toronto

This text is printed on acid-free paper.

Copyright © 1998 by John Wiley & Sons, Inc.

All rights reserved. Published simultaneously in Canada.

Reproduction or translation of any part of this work beyond
that permitted by Section 107 or 108 of the 1976 United
States Copyright Act without the permission of the copyright
owner is unlawful. Requests for permission or further
information should be addressed to the Permissions Department,
John Wiley & Sons, Inc., 605 Third Avenue, New York, NY
10158-0012.

This publication is designed to provide accurate and
authoritative information in regard to the subject
matter covered. It is sold with the understanding that
the publisher is not engaged in rendering legal, accounting,
or other professional services. If legal advice or other
expert assistance is required, the services of a competent
professional person should be sought.

Library of Congress Cataloging in Publication Data:

Logsdon, Tom, 1937–
      Orbital mechanics : theory and applications / Tom Logsdon.
          p.   cm.
      Includes bibliographical references and index.
      ISBN 0-471-14636-6 (cloth : alk. paper)
        1. Orbital mechanics.   I. Title.
   TL 1050.L59   1997
   521'.3—dc21   97-6507

10 9 8 7 6 5 4

This book is lovingly dedicated to Cyndy Logsdon, who deserves unlimited praise and accolades for all the helpful and considerate things she has done for me in making books, teaching courses, and in living life with significance. I was delighted when we got to end our long hours of hard work on the manuscript with an eagerly awaited working vacation in Strasbourg, France.

CONTENTS

Preface   xiii

Acknowledgments   xv

List of Symbols   xvii

1. An Introduction to Orbital Mechanics   1

   The Mystery of the Wandering Stars   2
      Ptolemy's Earth-Centered System   2
      The Heliocentric Theory of Copernicus   3
   Galileo's Critical Insights   7
   Tycho Brahe's Landmark Observations   8
   Kepler's Three Laws of Planetary Motion   9
   Isaac Newton's Clever Generalizations   11
      Newton's Universal Law of Gravitation   12
      Newton's Simple Laws of Motion   16
      Generalizing Kepler's Laws   17
   Early Pioneers in Orbital Mechanics   19
   Why Doesn't a Satellite Fall?   20
   Why Do Satellites Travel Around in Elliptical
      Orbits?   22
   The Conic Sections   23
   Launching a Satellite into Orbit   24
   Weightlessness   25

2. Satellite Orbits   29

   Isaac Newton's *Vis Viva* Equation   30
   Evaluating the Gravitational Parameter of Earth   32
   Escaping from the Surface of a Three-Mile
      Asteroid   34
   Elliptical Orbits   37
   The High Cost of Launching a Satellite   40

The Best Place to Escape from an Elliptical Orbit   42
Gravity Wells   44
Kepler's Equation   45
The Six Classical Keplerian Orbital Elements   53
Launch Azimuths and Ground-Trace Geometry   56
Orbital Perturbations   58

3. The Orbital Environment   63

The Beneficial Properties of Space   64
  No Gravity   64
  No Air   64
  Nothin' to Block Your View   65
  The Emerging Benefits of Space Technology
    to the Residents of Springfield, Kentucky   66
The Salient Characteristics of the Orbital
  Environment   66
Earth's Atmosphere   67
Earth's Gravitational Field   69
The Magnetic Field Surround Earth   71
Meteoroids in Space   74
Man-Man Space Debris   76

4. Powered Flight Maneuvers   81

The Classical Hohmann Transfer Maneuver   81
Low-Thrust and Multi-Impulse Maneuver
  Sequences   84
Pure Plane-Change Maneuvers   85
Combined Plane Changes   88
The Bielliptic Transfer   90
Walking-Orbit Maneuvers   92
On-Orbit Rendezvous   94
Deorbit Maneuvers   97
Space-Age Slingshots   101
Ballistic Capture Missions   103

5. Boosting a Satellite into Orbit   107

The Rocket as a Momentum Exchange Device   108
Robert Goddard's Contributions   109
Typical Liquid- and Solid-Fueled Rockets   111
Specific Impulse   113
The Rocket Equation   114

Multistage Rockets 116
    A Typical Single-Stage Rocket 116
    A Typical Two-Stage Rocket 117
    Trajectory Losses 118
Acceleration Profiles 120
Adding Lightness 121
    Balloon Tank Design 122
    Common Bulkheads 122
    Higher-Density Propellants 124
The High Cost of Accelerating Unburned
    Propellants 124
Staging Techniques 126

6. Today's Family of Global Boosters **127**

NASA's Reusable Space Shuttle 128
The Russian Space Shuttle 130
America's Expendable Rockets 132
    The Delta Booster 132
    The Atlas Family 133
    The Titan 134
The Winged Pegasus Booster 134
The European Ariane 136
Soviet Boosters on Parade 137
The Japanese H-2 139
The Chinese Long March 139
Advanced Booster Concepts Emerging from
    the Drawing Boards 139
    The X-33 Reusable Space Plane 139
    The Ariane V 140
    The Delta III 140
Novel Concepts for the Future 142
    Arthur Schnitt's Big, Dumb Booster 142
    Ed Keith's Asparagus-Stalk Booster 143
    The Maglifter Concept 146

**7. Enhancing the Performance of
Booster Rockets** **149**

Lunar-Orbit Rendezvous 150
Propellant Utilization Systems 151
The Programmed Mixture Ratio Scheme 153
Optimal Fuel Biasing 156

Optimal Trajectory Shaping   158
Postflight Trajectory Reconstruction   163
Summary of Performance Gains   166

8. Choosing the Proper Orbit for a Satellite   169

Polar Orbits and Polar Birdcage Constellations   169
   The Transit Navigation Satellites   170
   The Iridium Constellation   172
Sun-Synchronous Orbits   172
Full-sun, Sun-Synchronous Orbits   175
Geosynchronous Orbits   175
   Inclined Geosynchronous Orbits   175
   The Perturbations Acting on a Geosynchronous
     Satellite   177
   Nullifying the Effects of the Perturbations   182
12-Hour Semisynchronous Orbits   182
Russia's Molniya Communication Satellites   185
Ace-Orbit Constellations   185
Frozen Orbits   189
Three-Body, Libration-Point Orbits   190
Interplanetary Trajectories   192
Grand Tour Missions   195
Special Orbits in Review   196

**9. Choosing the Proper Constellation
Architecture   203**

What is a Constellation?   204
   A Sampling of Today's Constellations   204
   Spatial Locations for the Selected
     Constellations   207
What is the Largest Constellation Ever Launched
   into Space?   209
What is the Smallest Constellation of Satellites
   That Can Cover Earth?   210
   John Walker's Rosette Constellations   211
   John Draim's Four-Satellite Constellations   211
   The Smallest Constellation That Can Provide
     Global Coverage   213
Computer-Modeling Techniques   215
   Map Projections   216
   Coverage Analysis   217

The Space Eggs Computer-Simulation Program 218
Constellation Selection Trades for Mobile
   Communication Satellites 220
   The Orbital Environment 222
   The Estimated Cost of the Constellation 223
Cost and Complexity Assessments for the
   Overall System 227
The Global Benefits of Mobile Communication
   Satellite Constellations 229

**10. Space-Age Technologies for the
Twenty-First Century 231**

Chemical Mass Drivers 232
Electromagnetic Catapults 233
Nuclear Propulsion 234
   Project Orion 234
   The Nerva Rockets 234
Laser-Powered Rockets 236
Solar Sails 236
   The Soviet Regatta Satellites 237
   "Geostationary" Polesats 240
Tethered Satellites 240
Project Skyhook 242
   The Fundamentals of Space Exploration 242
   The Indian Rope Trick 243
   Tapering the Cable 243
   Dropping Satellites into Orbit 246
The Skyhook Pipeline 246
The Skyhook Complex 248

Bibliography 253

About the Author 263

Index 264

PREFACE

Even the simplest spaceborne maneuvers almost always produce counterintuitive results. If, for example, you are coasting around Earth in a 100 nautical-mile circular orbit, and you tap on the accelerator, your spaceship will immediately begin to slow down. Conversely, if you hit the brakes, it will speed up. And, it you attempt to chase down another spaceship sharing your orbit, dead ahead, by thrusting directly toward it, you will immediately begin to drift farther away.

Other seemingly some simple maneuvers in space produce even more bizarre results. Consider, for instance, the difficulties you will encounter in trying to get rid of unwanted garbage. If you throw a banana peel out the side window of your spaceship, 45 minutes later it will plunge back through that same window and slap you on the side of the face! If you throw it straight down through a hole in the floorboard, 90 minutes later, it will crash through the ceiling and hit you on the top of the head!

Counterintuitive paradoxes abound in orbital mechanics. But most popular books make only half-hearted attempts to explain these strange, unexpected results. Consequently, I felt compelled to assemble this slim volume to help professional practitioners and novices alike gain firm new insights into the curious movements of satellites and rockets traveling through space. In short, the purpose of this book is insight, not numbers.

In my quest to provide these most necessary insights, I have chosen to exploit a variety of fruitful analogies specifically designed to help you and your fellow readers understand and appreciate some of the more baffling aspects of orbital mechanics. Thus, in addition to the usual topics covered in standard works on the subject, I discuss several unusual concepts, including optimally contoured playground sliding boards, Mickey Mouse balloons filled with pink lemonade, well-executed mountain climbing expeditions, quarters spinning around in tight circles on drugstore counters, high-speed billiard balls bashing into one another, clever and productive "leaping lizards," and Skyhook Pipelines poking up through the stratosphere.

The ideas, insights, and applications emphasized in the various chapters of this book sprang naturally from years of joyful study and three full decades of practicing the craft. My quest to understand the intricacies of orbital mechanics started immediately after graduation at McDonnell Douglas in the spring of 1959, two years after the first Russian Sputnik shocked all of us out of our complacency. My first few projects at that ancient Santa Monica facility included powered flight maneuvers and performance evaluations for the Thor-Delta rocket and flight-film analyses for the Echo balloon.

Later, at Rockwell International, I was assigned to handle optimal trajectory shaping for the Apollo Moon flights, rendezvous and deorbit simulations for the manned Skylab, and systems analysis studies for the Navstar Global Positioning System (GPS).

With the world's greatest masterplan, I could never have mapped out a more exciting and enriching career. For nine years, I managed to eke out a living by putting men on the Moon, then I got to work on other projects that were equally demanding. Now I am delighted to invest a small part of my early retirement sharing a little of my hard-won expertise with the next generation of aerospace experts in person and on paper, too.

TOM LOGSDON

Seal Beach, California
November, 1997

ACKNOWLEDGMENTS

God: A being endowed with intelligence and wisdom.
—*World-ranking Bible authority Sir Isaac Newton*

A book is always a team effort and this one was no exception. Accordingly, I would like to take this opportunity to express my appreciation to the many talented individuals who helped me put this one together.

Over the past 15 years in 18 different countries scattered across 5 continents, my enthusiastic students, 5500 professional engineers and managers, have been a valuable resource in helping me structure, polish, and refine the contents of this book. Their many helpful contributions are enthusiastically acknowledged.

So are the patient and affectionate inputs from my perky soul mate, Cyndy, who handled most of the word processing for the various drafts of the manuscript. She was supported by Toma Nott, Anthony Vega, and Sandy Rimerez. Anthony and Sandy also helped design some of the figures and they transcribed most of the dictation tapes. Lloyd and Tinka Wing also used their trusty Macintosh computer to design many of the excellent drawings and graphs that accompany the text. The untiring efforts of these various individuals are much appreciated.

My friend Janis Indrikis checked many of the orbital mechanics derivations and calculations and he helped design a few of the tables and graphs. His efforts were supplemented by Jim Hafner and Alen Love who helped me straighten out some of the stickier technical tangles. Any errors that managed to elude them, however, are my sole responsibility. This responsibility is not being embraced with any measurable degree of enthusiasm. It's just not clear who else to blame.

Finally, I would like to thank Chad Arnett, a special child, who kept me awake and alert throughout the preparation of the manuscript. His boisterous "car noises" and his other thoughtful gestures of support much be acknowledged. So must his strange, hand-colored versions of some of the figures that accompany the text.

LIST OF SYMBOLS

ENGLISH SYMBOLS

a	Acceleration
a	Semi-major axis
A	Cross-sectional area
C_D	Drag coefficient
D	Drag force acting on a satellite
e	Orbital eccentricity
E	Eccentric anomaly
F	Force or thrust
F_c	Centrifugal force
F_g	Gravitational force
g	Gravitational acceleration
g_A	Surface gravitational acceleration of an asteroid
g_o	Gravitational acceleration at the surface of Earth
h	Angular momentum
h	Height or altitude
h_a	Apogee altitude
h_p	Perigee altitude
H	Scale height of Earth's atmosphere
i	Orbital inclination
I_{sp}	Specific impulse
J_2	The second potential harmonic of Earth's gravitational field (equatorial bulge term)
$L_1, \ldots, L_5$	Libration points 1 through 5
m	Mass
$\dot{m}$	Mass flow rate

M	Mass of Earth or some other celestial body
M	Mean anomaly
MR	Mixture ratio
n	Mean motion
P	Orbital period
P	Semi-latus rectum
P_e	Pressure of a rocket's exhaust gases when they escape from the end of its exhaust nozzle
P_E	Orbital period of Earth around the Sun
P_J	Orbital period of Jupiter around the Sun
P_L	Lapping interval (synoptic period)
P_o	Pressure of the ambient atmosphere
PL	Payload
r	Radius or radial distance
r_a	Apogee radius
R_A	Radius of an asteroid
R_E	Radius of Earth
R_E	Radial distance from the Sun to Earth
R_J	Radial distance from the Sun to Jupiter
r_o	Radius of Earth
r_p	Perigee radius
R_V	Radial distance from the Sun to Venus
t	Time
t_f	Final time
t_o	Initial time
T	Time of perigee (or perifocal passage)
V	Velocity
V_{CIRC}	Circular orbital velocity
V_e	Effective velocity of a rocket's exhaust molecules
V_{eA}	Escape velocity from the surface of an asteroid
V_{ESC}	Escape velocity
V_{ex}	Velocity of a rocket's exhaust molecules
V_{PL}	Velocity of the payload
w	Weight
$\dot{w}$	Flow rate
W_{ex}	Total weight of the exhaust of a rocket
W_f	Weight of a rocket at the end of its burn
W_o	Weight of a rocket at ignition

W_{PL}	Payload weight
W_{PROP}	Propellant weight
x, y, z	Rectilinear position coordinates
x_f, y_f, z_f	Final rectilinear position coordinates
x_o, y_o, z_o	Initial rectilinear position coordinates
$\dot{x}, \dot{y}, \dot{z}$	Mutually orthogonal velocity components
$\dot{x}_o, \dot{y}_o, \dot{z}_o$	Initial mutually orthogonal velocity components

GREEK SYMBOLS

α	Angle of attack. Current angle between a rocket's thrust vector and its current velocity vector
α	In-plane angle between the Sun and a specific planet as seen from Earth
α	Thrust angle
β	Solar beta angle. The angle between the current radius vector from the Sun and the orbit plane of a satellite
γ	Flight-path angle. Angle between the current velocity vector of a satellite and a line parallel to the local horizon
ΔE	Change in energy
ΔV	Change in velocity (or velocity increment)
ΔV_a	Velocity increment at apogee
ΔV_p	Velocity increment at perigee
ΔV_{TOT}	Total velocity increment
ΔW_{PROP}	Change in propellant weight (or weight of the incremental propellants burned)
θ	True anomaly. The Earth-centered angle between a satellite's current orbital position and that segment of the line of apsides of its orbit pointing toward perigee
$\Delta\theta$	Intersection angle between two satellite orbits or the change in that angle
$\Delta\theta_a$	Change in the intersection angle between two satellite orbits at apogee
$\Delta\theta_p$	Change in the intersection angle between two satellite orbits at perigee
μ	The gravitational parameter of a particular celestial body such as Earth or the Sun

ρ	Atmospheric density
ρ_{SL}	Atmospheric density at sea level
σ	Standard deviation (from statistics)
ω	Argument of perigee. The Earth-centered angle measured in the satellite's orbit plan between its ascending node and its current perigee location
ω	Rotation rate of any body
ω	Rotation rate of Earth
$\dot{\omega}$	Rotation rate of a satellite's line of apsides around the center of Earth (caused by external orbital perturbations)
Ω	Right ascension of the ascending node of a satellite's orbit. The angle measured in the equatorial plane between the vernal equinox and the satellite's current ascending node
$\dot{\Omega}$	Inertial nodal regression rate (rate of change in the location of the equatorial crossing point of a satellite's orbit)

1 AN INTRODUCTION TO ORBITAL MECHANICS

> Of all the mathematics developed up until the time of Isaac Newton, Newton's was by far the better half.—*Lifelong rival of Isaac Newton, Gottfried Wilhelm von Leibniz*

In October 1957, when the Russians launched their first Sputnik, I was a junior in college learning to enjoy mathematics and physics on a beautiful green campus in the Bluegrass region of Kentucky. Late that afternoon, I wandered down into the ravine, a green, grassy dimple in the middle of the campus that had long served as a combination outdoor amphitheater and lover's lane. Along the back row of the amphitheater, I found a perfect spot to ponder that new Russian satellite that, according to the newspapers, would continue to circle Earth for several hundred years.

"What keeps it up there?" I asked myself. "Why doesn't it fall?" All the heavy objects I had encountered—Spalding tennis balls, Campbell's soup cans, ballpoint pens—when released, inevitably fell down toward the ground. But somehow that new Russian Sputnik seemed to be defying the force of gravity that was tugging so relentlessly on everything else.

That baffling paradox occupied my thoughts for a few brief moments, then I shifted my attention to a much larger object ambling across that pale Kentucky sky. "The Moon doesn't fall," I suddenly realized. "Maybe that has something to do with it."

I had no way of knowing it at the time, but 290 years earlier, an energetic young undergraduate at Cambridge University had asked himself two strikingly similar questions: "Why does the Moon continue to sail over the British countryside? And why do apples and other loose objects inevitably plummet down toward the ground?"

1

His name was Isaac Newton, one of the brightest individuals who ever lived. Within a short time, he had formulated brilliant answers to those two seemingly simple questions. Then, armed with the answers, he quietly discovered the fundamental principles of orbital mechanics, an elegant branch of mathematical physics that helped Isaac Newton and his successors fathom the basic architecture of our solar system, the Milky Way Galaxy, and the universe beyond.

THE MYSTERY OF THE WANDERING STARS

Isaac Newton's creative insights can be traced back into the misty shadows of the past when ancient fishermen and shepherds noticed curious spiraling movements in the night sky. Most of the stars traveled from horizon to horizon along gently curling arcs always with essentially the same spacings. A few of them, however, followed strange looping trajectories with respect to their neighbors in space. The ancients called these erratic pinpoints of light *the wandering stars.*"

Figure 1.1 depicts the path of one of those wandering stars, the planet Venus, over a four-year period. Notice how it traces out a complicated path among the more distant background stars.

Why do the wandering stars, on occasion, appear to travel backwards across the night sky? The ancient astronomer, Ptolemy, who lived in Alexandria, Egypt, in the second century A.D., found a way to arrange the bodies in the solar system that seemed to account for their curious retrograde movements.

Ptolemy's Earth-Centered System

In Ptolemy's construction, planet Earth was, quite naturally, positioned at the center of the solar system. Orbiting around Earth were the seven wandering "stars" he had observed: Mercury, Venus, the Sun, the Moon, Mars, Jupiter, and Saturn. To a first approximation, each of those massive celestial bodies traveled along a precise circular arc. These circular movements were modified by so-called epicycles superimposed on more pronounced circular motions. An epicycle is like a wheel rolling on another wheel. Once Ptolemy had introduced enough wheels rolling on other wheels he was—at least to the crude accuracy of his celestial observations—able to duplicate the complicated motions of the wandering stars.

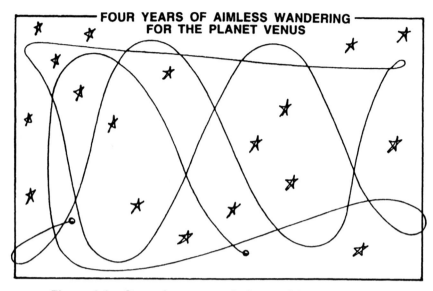

Figure 1.1 Over a four-year period, one of the wandering stars, the planet Venus, follows this strange, looping trajectory as seen against the much more distant background stars. Many natural philosophers attempted to explain why its path was so complicated, but more than 2000 years would go by before anyone could formulate an acceptable explanation for the planet's erratic behavior.

The Greek philosopher Aristotle enthusiastically embraced the geometrical constructions patched together by Ptolemy and, in turn, early Christian leaders convinced themselves that this was the only viable architecture for the solar system. That architecture also seemed to correlate well with their everyday observations. When they gazed up toward the night sky, the stars all appeared to move along circular arcs surrounding Earth. Moreover, the circle is a perfect geometrical shape uncorrupted by everyday bumps and squiggles, so it seemed to belong in the sacred heavens.

The Heliocentric Theory of Copernicus

In the sixteenth century, the Polish astronomer, Nicolaus Copernicus, decided that Ptolemy's architecture for the solar system did not provide the best explanation for the observed trajectories of the wandering stars. After studying the problem at length, Copernicus theorized the Sun was positioned at the center of our solar system. In his conception, Earth was merely one of the wandering stars traveling around the center of the Sun.

Unfortunately, the simple circular movements he postulated did not quite duplicate the complicated movements of the wandering stars. So he, too, had to superimpose epicycles on the more pronounced circular motions of the planets. Within a decade or two, he and his contemporaries had added dozens of epicycles to account for the subtle planetary movements they could observe.

Church leaders strongly resisted the theory of Copernicus. They reasoned that if they and their fellow Earthlings were riding around on a spinning planet, their bodies would be thrown off into space by centrifugal force. They were also convinced that because God's favorite creatures were so fundamentally important, Earth would necessarily occupy the center of the solar system.

Quantitatively at least, the architecture for the universe devised by Copernicus accounted for several of the mechanisms we observed in the sky. For instance, it explained the rising and setting of the Sun, Moon, and the stars. The gross motions of those celestial bodies was caused by the rotation of Earth around its north–south polar axis.

This new architecture also accounted for the waxing and the waning of the four seasons. The spin axis of Earth is tipped 23.5° with respect to the ecliptic plane—the plane that contains Earth and the Sun. Consequently, as Earth travels around the Sun, the Sun's perpendicular rays first illuminate the Northern Hemisphere (during our summer) and then the Southern Hemisphere (six months later, during our winter).

The architecture devised by Copernicus also explained the curious motion of the wandering stars. Figure 1.2 highlights the primary cause of their retrograde motions. Notice how Earth travels around the Sun in an orbit with a smaller radius than the orbit of Mars. Copernicus realized that when a planet is closer to the Sun, it travels at a faster angular rate. Consequently, as Figure 1.2 indicates, Earth sweeps past Mars to trace out an apparent retrograde loop.

Nicolaus Copernicus not only deduced the fundamental architecture of our solar system, he also figured out how big it is. His measurements were in astronomical units—the average distance between Earth and the Sun. He had no accurate way of estimating the magnitude of the astronomical unit, but we know today that is about 93 million miles.

The two diagrams in Figure 1.3 show how Copernicus determined the relative distances between each planet and the Sun. First, he worked out the orbital radii for the *interior planets*

THE HELIOCENTRIC THEORY OF COPERNICUS

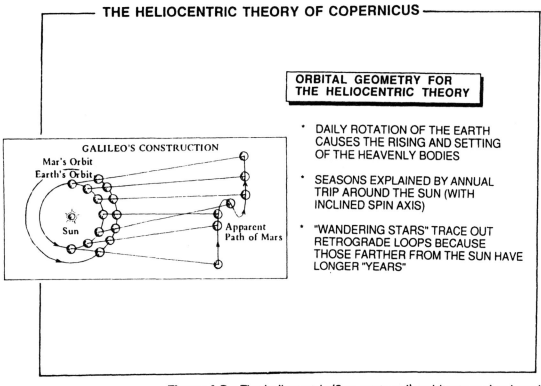

ORBITAL GEOMETRY FOR
THE HELIOCENTRIC THEORY

* DAILY ROTATION OF THE EARTH
  CAUSES THE RISING AND SETTING
  OF THE HEAVENLY BODIES

* SEASONS EXPLAINED BY ANNUAL
  TRIP AROUND THE SUN (WITH
  INCLINED SPIN AXIS)

* "WANDERING STARS" TRACE OUT
  RETROGRADE LOOPS BECAUSE
  THOSE FARTHER FROM THE SUN HAVE
  LONGER "YEARS"

GALILEO'S CONSTRUCTION

Mar's Orbit

Earth's Orbit

Sun

Apparent
Path of Mars

Figure 1.2 The heliocentric (Sun-centered) architecture developed by Nicolaus Copernicus represented a bold attempt to explain some of the things he and others had observed. These included the nightly motions of stars along the celestial sphere, the systematic reappearances of the four seasons, and the curious retrograde movements of the wandering stars as seen against the infinitely distant background stars.

(the ones whose orbits are inside the orbit of Earth). Then he used a slightly more complicated algorithm to work out the orbital radii for the *exterior planets* (the ones whose orbits are beyond the orbit of Earth.)

His method for calculating the distance from the Sun to the two interior planets, Mercury and Venus, is presented on the left-hand side of Figure 1.3. For this calculation, Copernicus waited for Venus (or Mercury) to reach its greatest angular distance from the Sun as seen from Earth. Then he constructed the right triangle sketched on the left-hand side of Figure 1.3. Once he had measured this maximum angle between the Sun and Venus, he could calculate the radius of the orbit of Venus from the following trigonometric equation:

$$R_V = R_E \sin \alpha$$

— HOW DID COPERNICUS DETERMINE THE RELATIVE SIZES
OF THE PLANETARY ORBITS?

☆ MERCURY
☆ VENUS

☆ MARS
☆ JUPITER
☆ SATURN

THE INTERIOR PLANETS

THE EXTERIOR PLANETS

LAPPING INTERVAL
GIVES ANGULAR
TRAVEL RATE

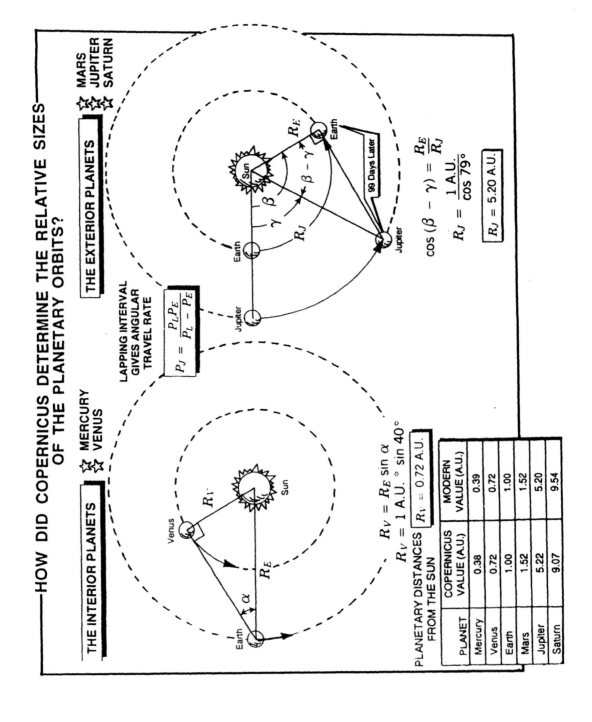

$$P_J = \frac{P_L P_E}{P_L - P_E}$$

$$\cos (\beta - \gamma) = \frac{R_E}{R_J}$$

$$R_J = \frac{1 \text{ A.U.}}{\cos 79°}$$

$$R_J = 5.20 \text{ A.U.}$$

99 Days Later

$$R_V = R_E \sin \alpha$$

$$R_V = 1 \text{ A.U.} \cdot \sin 40°$$

$$R_V = 0.72 \text{ A.U.}$$

PLANETARY DISTANCES
FROM THE SUN

PLANET	COPERNICUS VALUE (A.U.)	MODERN VALUE (A.U.)
Mercury	0.38	0.39
Venus	0.72	0.72
Earth	1.00	1.00
Mars	1.52	1.52
Jupiter	5.22	5.20
Saturn	9.07	9.54

where R_V is the radius of the orbit of Venus around the Sun, and R_E is one astronomical unit (Earth–Sun distance). For the exterior planets, Copernicus used a completely different algorithm. He started by determining the lapping interval (synoptic period) of Earth and Jupiter. Wait until Earth is directly between Jupiter and the Sun, and then measure the elapsed time until that same alignment occurs again. The desired alignment occurs approximately every 396 days, so about 99 days after such an alignment, Earth will be 90° ahead of Jupiter, as shown by the right-hand side of Figure 1.3. Simple trigonometry can then be used as shown in the figure to compute the distance between the Sun and Jupiter.

Copernicus' estimates for the radial distances between the Sun and the five planets that could be observed with the naked eye are presented in the lower left-hand corner of Figure 1.3. Copernicus was unable to estimate the size of the astronomical unit. But, nevertheless, he managed to work out the gross architecture of the solar system. His planet–Sun distances are surprisingly close to the ones we use today—which are also listed in the figure.

GALILEO'S CRITICAL INSIGHTS

In the fifteenth and the sixteenth centuries, the Italian physicist, Galileo Galilei, made a series of penetrating observations that later helped Isaac Newton in his quest to explain why the apples on his mother's apple tree inevitably tumbled down toward the ground whereas the Moon continued to sail overhead.

One morning in 1581, so the story goes, Galileo wandered into the church at the Leaning Tower of Pisa where he saw a lamp suspended from the ceiling on a chain gently swaying back and forth in the breeze. Hundreds of Italian churchgoers had seen that same lamp moving to and fro. But Galileo noticed something that none of them had ever noticed before. He

Figure 1.3 Using a clever line of reasoning, but only the simplest trigonometric calculations, Nicolaus Copernicus developed a satisfying architecture for the solar system. He also estimated the planetary distances in astronomical units (the distance separating Earth from the Sun). The planet–Sun distances he derived are tabulated in the lower left-hand corner. Notice how well his estimates compare with our modern values.

realized that the period of time required for the pendulum to travel back and forth was always the same regardless of the length of its swinging arc. If it traveled on a long arc, it moved more rapidly. If it traveled on a shorter arc, it moved more slowly to compensate.

Galileo never constructed any clocks based on the pendulum principle. But others did so and soon clock-making technology was sweeping across the European continent. Indeed, some historians credit Galileo's simple observation with the spawning of the Industrial Revolution. In their view, the science-based technologies used in building pendulum clocks later became the basis for the machinery that fostered rampant industrialization.

In 1592, Galileo conducted a series of experiments proving conclusively that heavy objects and light objects fall under the influence of gravity at the same constantly accelerating rate (neglecting drag with the atmosphere). This observation ran counter to the conjectures of Aristotle, who was convinced that big, heavy objects fall faster.

By 1604, Galileo had measured the rate at which falling objects accelerate and, as a result, he formulated his famous Law of Uniform Acceleration, which states that falling objects near the surface of Earth accelerate uniformly when they are pulled downward by the force of gravity. Within one second after it is released, an object falls 16 feet. In two seconds, it falls a total of 64 feet. And in the first five seconds after it is released, it falls 400 feet.

The Italian government has honored Galileo by placing his picture on their 2000-lira banknote—roughly $2 in U.S. currency. The United States is unquestionably today's preeminent power in science and technology, but only politicians—never scientists or engineers—are privileged enough to get their likenesses on U.S. money.

TYCHO BRAHE'S LANDMARK OBSERVATIONS

The sixteenth-century Danish astronomer Tycho Brahe compiled a series of precise observations that further aided Isaac Newton in formulating his laws of orbital mechanics. In 1576, Brahe constructed an astronomical observatory on an island off the coast of Denmark to house the ungainly, but precise instruments he used in observing the celestial objects in the night sky. Some of his instruments were as big as houses, but they had no contoured mirrors or lenses to provide magnification of the celestial objects under study. They were, in effect, oversized protractors fashioned to allow Brahe to measure angles with superb precision. Two thousand years earlier, the ancient Greeks had constructed conceptually similar instruments, but their versions were much smaller and much less accurate.

For nearly 20 years, Tycho Brahe used his enormous instruments to determine and record the angular positions of the five visible planets. He also measured the locations of the more distant background stars with unprecedented precision. Where necessary, he corrected his observations for atmospheric refraction. This he accomplished by noting that when the distant stars were near the horizon, their observed positions were shifted slightly.

The earth's atmosphere creates surprisingly large refractions. When lovebirds are observing the setting Sun with its lower edge just tangent to the horizon, the entire Sun is actually situated below the horizon. If the atmosphere suddenly disappeared, the setting Sun would not be visible at all!

Tycho Brahe recorded his observations with painstaking care. As he grew older, his assistant, Johannes Kepler, tried desperately to get his hands on those amazingly detailed and accurate observations. On his deathbed, Brahe finally released the tables to his young assistant who suddenly had access to a treasure trove of information from which he was, not without difficulty, able to devise his three well-known laws of planetary motion.

KEPLER'S THREE LAWS OF PLANETARY MOTION

Armed with the laboratory notebooks of Tycho Brahe, Johannes Kepler spent several years trying to figure out exactly how the wandering stars travel around the Sun. He knew that the Sun-centered architecture of Nicolaus Copernicus provided a reasonably satisfactory explanation. But when that architecture was tested with sufficiently precise observations, it quickly began to fall apart.

At one point, Kepler theorized that the planets might be tracing out circular orbits with centers that were offset with respect to the center of the Sun. This, however, did not turn out to be the correct answer, so he had to start all over again.

Johannes Kepler's fundamental difficulty was that Tycho Brahe's observations were not made from some omnipotent vantage point deep in outer space. Instead, he had made them while riding on a rapidly moving platform called Earth. Consequently, Kepler faced two separate, but intertwined problems. First, he had to figure out what path Earth followed as it traveled around the Sun. Then he had to figure out what paths the planets followed in making their separate journeys.

This problem is analogous to the one that might be faced by an observant scientist who is riding on a merry-go-round at night while observing a small light attached to a moving Ferris wheel. The scientist, who sees the combined effects of both movements simultaneously, must somehow separate the individual movements of the merry-go-round from those of the Ferris wheel.

Johannes Kepler developed a surprisingly clever technique for separating the motion of Earth from the motions of the other five visible planets. In the process, he devised an entirely new architecture for the solar system. At one point, his patience was severely tested when he spent four agonizing years calculating the orbit of Mars using a precise, but very time-consuming mathematical algorithm. Unfortunately, no matter how many adjustments he made, circular orbits could not be made to match the planetary positions Tycho Brahe had observed. Eventually, Kepler abandoned circular orbits and tried, instead, a different type of orbit with an oblong shape that had been studied by ancient Greek mathematicians. Soon he found that the orbit of Mars fits Brahe's observations with great precision if he assumed that Mars was tracing out an *elliptical orbit*. This led him to formulate the first of his three famous laws of planetary motion, all of which are listed in Figure 1.4.

Kepler's first law states that every planet in the solar system moves around the Sun along an elliptical orbit with the sun at one focus. His second law defines the ever-changing motion of each planet hurtling around the Sun. It states that the line joining the Sun and the planet sweeps out equal areas in equal times. Connect the Sun and the planet with a straight line called the *radius vector*. As the planet travels around the sun, both its velocity and its radius vector are constantly changing so that the radius vector sweeps out equal areas in equal times. Each of the two shaded regions in the middle of Figure 1.4 represents one month of travel for the planet Mars. Notice that when the planet moves in closer to the Sun, it speeds up just enough to compensate for the fact that, at that closer range, its radius vector is shorter.

Kepler's third law states the squares of the periods of the planets are proportional to the cubes of their mean distances

KEPLER'S THREE LAWS OF ORBITAL MOTION

KEPLER'S FIRST LAW

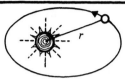

"EACH PLANET MOVES ALONG AN ELLIPTICAL ORBIT WITH THE SUN AT ONE FOCUS."

$$r = \frac{a(1 - e^2)}{1 + e \cos \theta}$$

KEPLER'S SECOND LAW

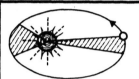

"THE LINE JOINING THE SUN AND THE PLANET SWEEPS OUT EQUAL AREAS IN EQUAL TIMES."

$$h = r^2 \frac{d\theta}{dt} = \text{Constant}$$

KEPLER'S THIRD LAW

"THE SQUARES OF THE PERIODS OF THE PLANETS ARE PROPORTIONAL TO THE CUBES OF THEIR MEAN DISTANCES FROM THE SUN."

$$P = \frac{2\pi}{\mu^{1/2}} a^{3/2}$$

Figure 1.4 Johannes Kepler carefully analyzed the planetary positions compiled by Tycho Brahe in deriving these three simple laws of planetary motion. Kepler's first law defines the shape of the planetary orbits. His second law pinpoints the changing velocity at which the planets travel around the Sun. His third law provides us with an accurate estimate of the total amount of time required for each of the planets to complete one orbit.

from the Sun. If a planet is at a greater average distance from the Sun, it has farther to go to complete one 360° circuit, but it also travels at a slower angular rate. It is the combined effect of the extra path length and the lower angular rate of travel that creates the 3/2 power law linking the semi-major axis and the orbital period (see Figure 1.4).

ISAAC NEWTON'S CLEVER GENERALIZATIONS

Isaac Newton, Christmas present to the world, was born on Christmas Day in 1642, the same year Galileo died. His father died before he was born, and he was a tiny premature baby. His mother later observed that his entire body would have fit inside a quart beer mug. At birth, Newton's neck was so weak a doctor at Woolsthorpe made him a bolster—a small neck brace—to support the weight of his head.

In his formative years, Isaac Newton was not an especially good student, but he had creative ideas and was very clever with his hands. He built a wooden doll house and a little windmill backed up by one mousepower. When the wind refused to blow, the mouse would run inside a rotating cylinder to provide the necessary motive power. The young Newton also constructed a kite that carried a lantern over the countryside at Woolsthorpe, thus, perhaps, creating one of the earliest UFOs ever observed. In these days, ordinary people were definitely not accustomed to seeing lights wobbling up and down as they traveled across the night sky.

Newton's Universal Law of Gravitation

In 1665, when Isaac Newton was an undergraduate at Cambridge University, the second Great Plague raged across the British Isles and Cambridge was shut down. Newton then returned to his boyhood home at Woolsthorpe. There, according to his own account, he noticed an apple falling from a tree. That simple observation caused him to challenge himself with a powerful question: "Why doesn't the moon also fall down toward the ground?"

Earlier in that same year, Newton had been experimenting with beams of light, and he knew from those experiments that if we move twice as far away from a light source, a unit area will intercept one-fourth as much radiant energy. In other words, its intensity per unit area follows an inverse square law of intensity. Newton did not know why, but he conjectured that the force of gravity might behave in a similar way, that is, its strength would be inversely proportional to the square of the distance between some orbiting object under study and the center of Earth.

Newton knew that the Moon was roughly 60 times as far away from the center of Earth as his mother's apple tree. Thus, in accordance with his conjecture, the gravitational acceleration it would experience would be weaker by a factor of 3600. At that lower rate of acceleration, the Moon would fall down toward Earth only 16 feet in one minute, as shown in Figure 1.5.

The Moon travels around the center of Earth at 2000 miles per hour. So in one minute, it travels 38 miles along a trajectory perpendicular to the radius vector connecting it to the center

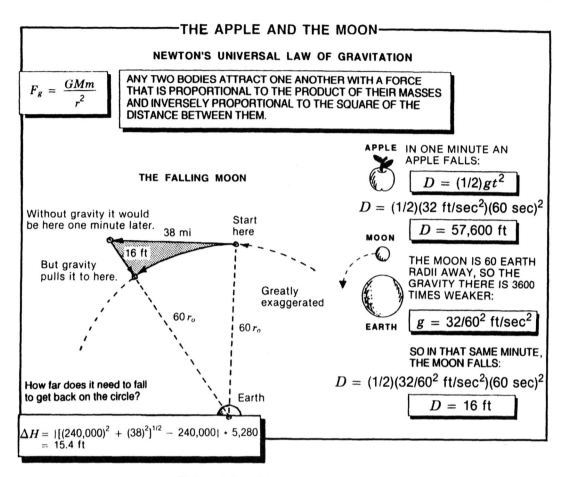

Figure 1.5 Using the law of uniform acceleration first devised by Galileo, Isaac Newton formulated a brilliant thought experiment to demonstrate that the observed motion of the Moon is compatible with his conjectured inverse square gravitational law. Newton verified this conjecture by calculating where the Moon would end up if it fell under gravity for one minute while traveling the same amount of time at its observed 2000-mph tangential velocity. The Pythagorean theorem then allowed him to demonstrate that these two combined motions would cause the Moon to fall back into the same circular orbit over and over again.

of Earth. Of course, during that same minute, the Moon is pulled down toward Earth 16 feet. As the calculation at the bottom of Figure 1.5 indicates, a fall of 16 feet is just sufficient to put the Moon back on its original circular orbit. Consequently, as Newton correctly concluded, the Moon does fall—it falls continuously around Earth along a gently curving circular trajectory.

From these deceptively simple calculations, Newton arrived at a sweeping conclusion that can be stated as follows:

Every particle in the universe attracts every other particle with the force that is directly proportional to the product of their masses and inversely proportional to the distance between their centers.

Unfortunately, Newton enountered great difficulties when he tried to use this general rule to establish how large bodies such as planets and their moons attract other nearby objects. He conjectured correctly that, if a planet is composed of concentric spherical shells of equal-density material, it will attract other objects in the universe as though all of its mass is concentrated at its center. Eventually, he was able to use integral calculus, which he invented, to prove that this conjecture is correct. That powerful theorem greatly simplifies many orbital mechanics calculations. A simple corollary tells us that if an ambitious coal miner tunnels down inside Earth, all the material that is at greater radial distance from Earth's center than his body will cancel out gravitationally. Thus, as he gets closer and closer to the center of Earth, his body will become lighter and lighter. When he finally arrives at its center, he will be weightless.

If we assume that Earth is of uniform density throughout, the gravitational acceleration the miner will experience inside Earth follows a direct power law:

$$g = \frac{g_0 r}{r_0}$$

where r_0 is the radius of the earth, g_0 is the surface acceleration (32 ft/sec^2), and g is the local gravitational acceleration at any radius r from the center of Earth.

Above Earth, of course, the gravitational acceleration follows an inverse square law:

$$g = \frac{g_0 r_0^2}{r^2}$$

where the g_0, g, r_0, and r have the same meanings as in the preceding equation.

The graph in Figure 1.6 defines the gravitational accelerations objects experience at various altitudes above Earth. Unfortunately, we live at the peak of this sharp curve. Small wonder so many of us spend our lives on difficult weight-reduction diets. If we lived inside Earth—or high above it—we would weigh considerably less than we do as we crawl around on Earth's surface.

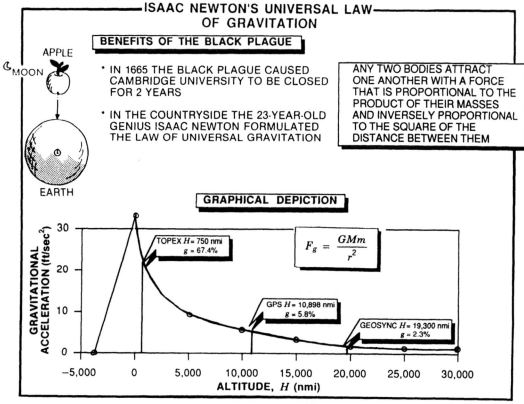

ISAAC NEWTON'S UNIVERSAL LAW OF GRAVITATION

APPLE

$\mathfrak{C}_{MOON}$

EARTH

BENEFITS OF THE BLACK PLAGUE

* IN 1665 THE BLACK PLAGUE CAUSED CAMBRIDGE UNIVERSITY TO BE CLOSED FOR 2 YEARS

* IN THE COUNTRYSIDE THE 23-YEAR-OLD GENIUS ISAAC NEWTON FORMULATED THE LAW OF UNIVERSAL GRAVITATION

ANY TWO BODIES ATTRACT ONE ANOTHER WITH A FORCE THAT IS PROPORTIONAL TO THE PRODUCT OF THEIR MASSES AND INVERSELY PROPORTIONAL TO THE SQUARE OF THE DISTANCE BETWEEN THEM

GRAPHICAL DEPICTION

$$F_g = \frac{GMm}{r^2}$$

TOPEX $H = 750$ nmi
$g = 67.4\%$

GPS $H = 10,898$ nmi
$g = 5.8\%$

GEOSYNC $H = 19,300$ nmi
$g = 2.3\%$

GRAVITATIONAL ACCELERATION (ft/sec^2)

ALTITUDE, H (nmi)

Figure 1.6 When the first Russian Sputnik was hurled into space, some early newspaper accounts claimed that it was not pulled back down again because it was high above Earth's gravitational field. However, even at the geosynchronous altitude 19,300 nautical miles high, Earth's gravitational force is 2.3 percent as strong as it is here at Earth's surface. At the Moon's distance 200,000 nautical miles away, Earth exerts a gravitational force 0.03 percent as strong as we Earthling's experience in our daily lives.

When the Russians launched their first Sputnik into orbit, some U.S. newspapers reported that it did not fall back because once it reached its destination altitude, it was completely above the gravitational influence of Earth. However, without gravity, Sputnik would not have remained in orbit. Actually, as Figure 1.6 indicates, the force of gravity Earth exerts on a low-altitude satellite is only slightly weaker than the gravitational force we routinely experience here on its surface.

Newton's Simple Laws of Motion

In addition to his universal law of gravitation, Isaac Newton also formulated three other important laws governing the behavior of moving objects. His three laws of motion can be stated quite simply (see Figure 1.7).

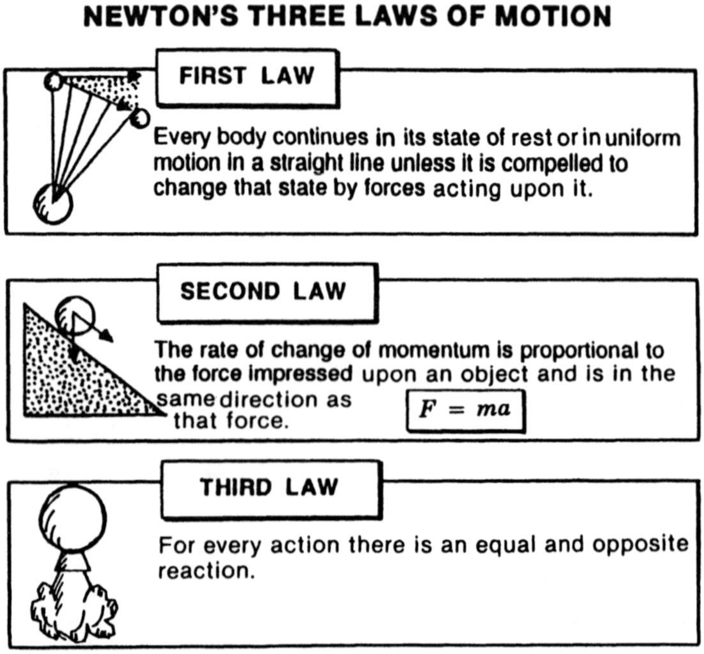

NEWTON'S THREE LAWS OF MOTION

FIRST LAW

Every body continues in its state of rest or in uniform motion in a straight line unless it is compelled to change that state by forces acting upon it.

SECOND LAW

The rate of change of momentum is proportional to the force impressed upon an object and is in the same direction as that force.

$$F = ma$$

THIRD LAW

For every action there is an equal and opposite reaction.

Figure 1.7 In his landmark publication, *Principia*, Isaac Newton formulated these three simple laws that govern the manner in which objects move. Contrary to intuition, his first law states that a body will continue to move forward forever at a uniform rate in the absence of external forces. Newton's second law provides us with a measure of the rate of acceleration of a body in terms of the force being exerted on it and its mass. His third law indicates that every action is accompanied by an equal and opposite reaction. Taken together, these three laws and the universal law of gravitation form the conceptual basis for most of orbital mechanics.

Newton's first law states: *Every body continues in its state of rest or in uniform motion in a straight line unless it is compelled to change that state by forces acting upon it.* Two thousand years earlier, Aristotle convinced himself that the "natural" state of objects was motionlessness. In his view, a body could continue to move only if it had a force acting on it at all times.

By contrast, Isaac Newton realized that it is no more natural for a body to remain still than for it to move in a straight line forever. This penetrating insight is all the more amazing when we pause to consider that Newton hardly ever saw any object move in a straight line for any length of time. Perhaps he skipped a flat stone on a frozen lake and watched it as it gradually slowed down. Otherwise, he had little opportunity to make direct observations supporting his first law.

Newton's second law (see Figure 1.7) states: *The rate of change of momentum is proportional to the force impressed upon an object and is in the same direction as that force.* Often, we simplify the second law by writing it as follows:

$$F = ma$$

where m is the mass of the body, F is the force impressed upon it, and a is the acceleration it experiences as a result. This law, too, is deceptively simple. It involves technical terms that we hear quite frequently—force, mass, and acceleration. Isaac Newton, however, had to devise the law while he was figuring out precisely what was meant by all of these technical terms. This was no easy accomplishment, particularly in an era when words were used in very loose ways even by accomplished scientists.

Newton's third law states: *For every action there is an equal and opposite reaction.* This law is especially useful in figuring out how a rocket hurls its payload into space. When the rocket ejects exhaust molecules rearward, the reaction force hurls the rocket itself in the opposite direction. Thus, a rocket can be regarded as a momentum-exchange device.

Generalizing Kepler's Laws

By appealing to these three fundamental laws of nature together with his universal law of gravitation, Newton was able to prove that all three of Kepler's laws were correct. Moreover, in the process, he generalized all three of them! Kepler's first law, for instance, states: *Every planet moves along an elliptical orbit with the sun at one focus.* Newton showed that if we give a small satellite the proper velocity, it will coast around a large spherical planet (such as Earth) along an elliptical orbit.

However, he also showed that, depending on its initial position and velocity, it might, instead, trace out a parabola, a hyperbola, a circle, even, in some cases, a straight line. Newton also modified Kepler's first law by noting that if both of the two bodies in question have appreciable mass, the smaller body will not orbit about the center of the larger body. Instead, both of them will orbit around their common barycenter. A similar phenomenon can be observed at a football game. When a majorette tosses her baton into the air, it does not rotate around the heavy end. Instead, the entire baton rotates about its center of mass.

The Moon is a big, heavy satellite with an appreciable mass compared with the mass of Earth. Consequently, both the Moon and Earth orbit about their common barycenter. That common barycenter is inside Earth, but not at Earth's center. Instead, it lies along the line connecting Earth and the Moon, roughly two-thirds of the way from the center of Earth to its surface.

Kepler's second law states: *The radius vector joining the center of the sun and the center of a planet sweeps out equal areas in equal times.* Newton proved that this law is correct and that it is a direct consequence of the conservation of angular momentum. He also showed that it is a general law that holds for all *central force fields* no matter what law of attraction (or repulsion) they are following.

Kepler's third law states: *The squares of the periods of the planets are proportional to the cubes of their mean distances from the Sun.* Isaac Newton proved mathematically that this law accurately characterizes the periods of the planets. But, in addition, he showed that if the two bodies have appreciable mass, this simple law governing the orbital period must be modified to include a new constant of proportionality together with the sum of their masses in the denominator. This modification is necessary because both massive bodies orbit about their common barycenter.

EARLY PIONEERS IN ORBITAL MECHANICS

Figure 1.8 highlights the life spans of Copernicus, Galileo, Tycho Brahe, Johannes Kepler, and Isaac Newton. Newton never met any of the other four individuals whose life spans are

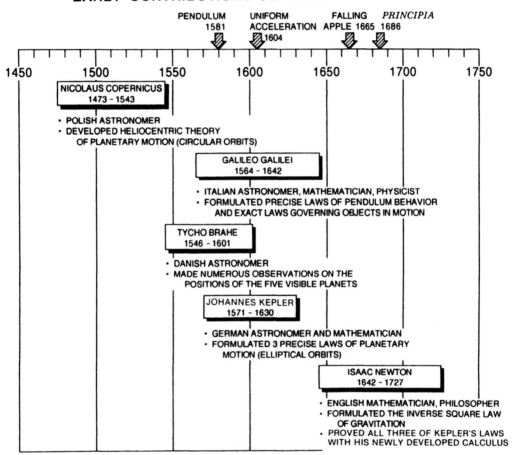

Figure 1.8 Isaac Newton made excellent use of the work of Nicolaus Copernicus, Galileo Galilei, Tycho Brahe, and Johannes Kepler, but he never met any of those important individuals. All of them were dead by the time Newton was born, so he knew them only through their reputations and their publications.

portrayed in Figure 1.8. He was born in 1642, the year Galileo died. By then Copernicus, Brahe, and Kepler were already dead.

In my three-day short courses on orbital mechanics, I ask my students a series of questions to make sure they are still alive—and awake! The questions are all different, but all the answers are the same. The answer to every question I ask them is "Isaac Newton."

WHY DOESN'T A SATELLITE FALL?

Why doesn't an orbiting satellite fall down to the ground? As Figure 1.9 indicates, it is the high speed of a satellite that keeps it in orbit. The small boy at the top of the figure has tied a stone to the end of a string and he is whirling it around in a tight circle. This simple drawing provides us with a beautiful analogy for the behavior of an orbiting satellite.

As the diagram at the bottom of Figure 1.9 indicates, if a satellite is to remain in a circular orbit, the centrifugal force pulling it upward must equal the gravitational force pulling it down toward the center of Earth. The centrifugal force is given by

$$F_c = mr\omega^2$$

where m is the mass of the satellite, r is its radial distance from the center of Earth, and ω is the angular rate at which it is traveling around its circular orbit.

At that same radius r, the force of gravity is given by

$$F_g = \frac{m\mu}{r^2}$$

where μ is the gravitational parameter of Earth. For circular orbits the centrifugal force and the gravitational force must equal one another. Otherwise, the satellite would move outward away from the center of Earth or closer to its center.

By equating the two forces, we can solve for the orbital velocity V at any given radius r:

$$V = \left(\frac{\mu}{r}\right)^{1/2}$$

WHY DOESN'T A SATELLITE FALL?

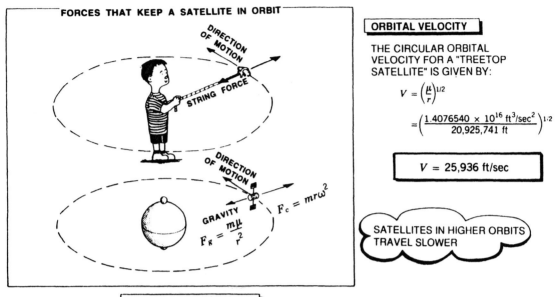

FORCES THAT KEEP A SATELLITE IN ORBIT

DIRECTION OF MOTION

STRING FORCE

DIRECTION OF MOTION

GRAVITY

$$F_g = \frac{m\mu}{r^2}$$

$$F_c = mr\omega^2$$

ORBITAL VELOCITY

THE CIRCULAR ORBITAL VELOCITY FOR A "TREETOP SATELLITE" IS GIVEN BY:

$$V = \left(\frac{\mu}{r}\right)^{1/2}$$

$$= \left(\frac{1.4076540 \times 10^{16} \text{ ft}^3/\text{sec}^2}{20,925,741 \text{ ft}}\right)^{1/2}$$

$$V = 25,936 \text{ ft/sec}$$

SATELLITES IN HIGHER ORBITS TRAVEL SLOWER

THE VELOCITY EQUATION

FOR A CIRCULAR ORBIT, F_c MUST COUNTERBALANCE F_g SO

$$\frac{m\mu}{r^2} = mr\omega^2 \longrightarrow r^2\omega^2 = V^2 = \frac{\mu}{r}$$

AND

$$V = \left(\frac{\mu}{r}\right)^{1/2}$$

Figure 1.9 A satellite can remain in orbit only because of its high rate of speed. This simple calculation uses a balance between local gravity and centrifugal force to show that the velocity needed to keep a low-altitude "treetop" satellite in orbit is nearly 26,000 feet per second, or 18,000 miles per hour.

Notice that the orbital velocity of a satellite is independent of its mass. Galileo's experiments demonstrated that heavy and light objects fall at the same rate of acceleration near Earth's surface. This equation shows that they would also "fall" around their orbits at the same rate.

The simple mathematical calculation at the top of Figure 1.9 uses the velocity equation we have just derived to determine the circular orbital velocity of a so-called "treetop satellite." A satellite traveling around a circular "treetop orbit" barely skims over the surface of Earth. Its required velocity turns out to be 25,936 feet per second, or about 18,000 miles per hour.

A few years ago, I demonstrated this concept to my stepson, Chad, who was fascinated by the whirling stone. When I finished the demonstration, I told him that a spaceship can remain in orbit only if it travels at an extremely high speed. "Star Trek doesn't," he replied.

WHY DO SATELLITES TRAVEL AROUND IN ELLIPTICAL ORBITS?

What causes a satellite to trace out an elliptical orbit? Several years ago, quite by accident, my daughter, Donna, and I stumbled on a terrific analogy for elliptical orbits. We were strolling across the campus at Warren High School in Downey, California, when we encountered two high school students playing a strangely informative game called tetherball.

In tetherball, a soft rubber ball is attached to the top of the pole with a flexible cord. One of the participants hits the ball with his hand in an attempt to wind the string around the pole in the clockwise direction. The other attempts to wind it in the counterclockwise direction. My daughter, who was only seven or eight years old at the time, courageously walked over to the young athletes and asked if she could join their game.

While I was watching them play, I realized that the moving ball closely mimics the behavior of a satellite traveling around an elliptical orbit. When the ball is struck with great force, it travels around a tilted, circular trajectory first climbing up against gravity, then falling back down again. Kepler's law of equal areas is, in effect, concealed within the ball's systematic movements. As it climbs upward against gravity, it gradually slows down until it reaches the apex of its flight. Then it falls back down to a lower height, picking up additional speed on the descending portion of its curved trajectory.

As the ball coasts around and around the top of the pole, it is constantly making a trade-off between potential energy and kinetic energy. Neglecting friction, it slows down, then speeds up again so that the sum of its kinetic energy and its potential energy is always constant.

When the game was over, Donna came running back toward me. "He won me," she said. "I knew he would. But I wanted to play." All of us, I suddenly realized, constantly find ourselves in that poignant situation. When we venture to Monte Carlo or Las Vegas, we harbor no illusions. We know they will win us. But we want to play.

THE CONIC SECTIONS

Ancient Greek mathematicians had understood and appreciated many of the mathematical properties of conic sections for more than 2000 years before Johannes Kepler and Isaac Newton figured out how these simple shapes accurately characterize the motions of celestial bodies.

As Figure 1.10 indicates, the conic sections can be created by slicing a cardboard dunce cap at different angles with a big meat cleaver. Notice how the various slices result in circles,

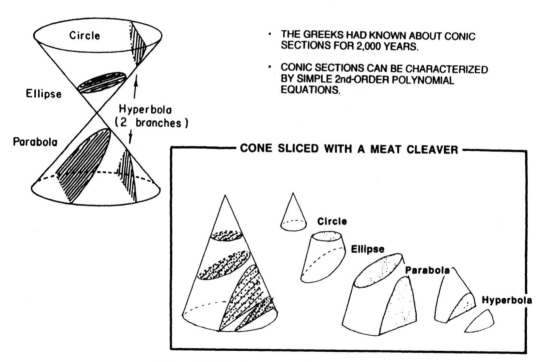

THE CONIC SECTIONS

- THE GREEKS HAD KNOWN ABOUT CONIC SECTIONS FOR 2,000 YEARS.

- CONIC SECTIONS CAN BE CHARACTERIZED BY SIMPLE 2nd-ORDER POLYNOMIAL EQUATIONS.

Figure 1.10 Ancient Greek mathematicians teased out many of the interesting properties of the conic sections 2000 years before Isaac Newton used them in his mathematical derivations. The conic sections can be created by slicing a dunce cap at various angles, thus producing circles, ellipses, parabolas, hyperbolas, and, in some cases, simple straight lines. The conic sections are represented by extremely simple mathematical equations. All of them can be expressed as second-order polynomials in the rectilinear coordinates x and y.

ellipses, parabolas, hyperbolas, and even, in some cases, straight lines. The conic sections are all defined by second-order polynomials expressed in rectilinear x and y coordinates. Second-order polynomials are the simplest equations defining curves that can close on themselves.

LAUNCHING A SATELLITE INTO ORBIT

When he was only 23 yeas old, Isaac Newton described, with remarkable clarity, how the Moon remains in orbit around Earth. Twenty-two years later in his book *System of the World*, he conjured up a "thought experiment" that explains how we can launch satellites into a circular or an elliptical orbit. In that thought experiment, he sent a strong athlete to the top of a mountain who threw stones horizontally at ever increasing speeds.

As Figure 1.11 indicates, the faster a stone is thrown, the farther it will travel before it impacts Earth. When such a stone reaches a speed of 18,000 miles an hour, it will not fall at all. Instead, it will enter a permanent orbit around Earth: Ninety minutes later, it will travel all the way around the world and hit the young athlete in the back of the head!

Thus, we now understand, as Isaac Newton did, that a satellite is held in orbit by its high rate of speed, which must be accurately controlled. When our country's aerospace engineers launched the Mercury astronauts into orbit, they had to control their orbital speed within half a percent.

If the astronauts had entered their orbit a half a percent *too fast*, their spacecraft would have carried them into an elliptical orbit that would have pierced the lower Van Allen Radiation Belt. If they had entered orbit a half a percent *too slow*, their spacecraft would have reentered Earth's atmosphere and fallen back to the ground at some uncontrolled location.

In the early days of the U.S. Space Program, when my friend Bob Africano and I were trying to understand the complexities of orbital mechanics, we developed a simple slogan intended to focus our attention on what really matters: *In space, velocity is the name of the game.* As you will undoubtedly notice in subsequent chapters of this book, we almost always control the motion of a space vehicle by carefully controlling its velocity.

ISAAC NEWTON EXPLAINS HOW TO LAUNCH A SATELLITE

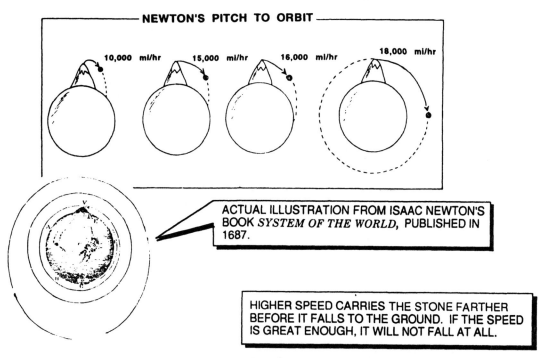

NEWTON'S PITCH TO ORBIT

ACTUAL ILLUSTRATION FROM ISAAC NEWTON'S BOOK *SYSTEM OF THE WORLD*, PUBLISHED IN 1687.

HIGHER SPEED CARRIES THE STONE FARTHER BEFORE IT FALLS TO THE GROUND. IF THE SPEED IS GREAT ENOUGH, IT WILL NOT FALL AT ALL.

Figure 1.11 In 1687, Isaac Newton published a book called *System of the World* in which he described how a satellite can be launched from a mountaintop into various circular or elliptical orbits. His original sketch, which is presented at the bottom of this figure, shows what happens when a young athlete throws stones off the top of a mountain at ever-increasing speeds. The faster a stone is traveling, the farther it will go before it slams into the ground. If it reaches a speed of 18,000 miles per hour, it will swing into a low-altitude circular orbit.

WEIGHTLESSNESS

When the first few Russian cosmonauts and American astronauts were launched into space, newspaper reporters of the day desperately tried to explain why they were experiencing weightlessness. One often repeated, but completely erroneous, explanation was that their powerful booster rockets had carried them completely above Earth's gravitational field.

Never mind that their space capsules would not stay in orbit at all if they did not have gravity to hold them in place.

Today, of course, we know that our astronauts experience the sensation of weightlessness because their bodies, the objects floating around inside their space capsules, and the capsules themselves all trace out free-fall trajectories around Earth.

When I was a young boy growing up wild and free in the Bluegrass region of Kentucky, my brother described weightlessness to me long before the space program had begun. According to his explanation: "If you step into an elevator shaft and turn loose of your car keys as you fall, they will just hang off your fingertips."[1]

The reason you will experience weightlessness as you are tumbling down an elevator shaft is that both your body and your car keys are accelerated downward at the same rate. The space shuttle astronauts who enjoy gulping free-floating shrimp from midair are experiencing exactly the same thing. Their bodies, their shrimp, and their space capsule are all falling along the same free-fall trajectories at exactly the same rate.

Weightlessness does not require a simple inverse square gravitational field as provided by a single, large celestial body. If 8 or 10 nearby planets are pulling on your body, your shrimp, and your space capsule, simultaneously, you would still be experiencing weightlessness—assuming that you are coasting along a free-fall trajectory with no forces to resist your fall. Here on Earth, we feel weight because our feet are touching the ground. Pull the ground out from under a menacing enemy and he will instantly experience weightlessness.

Orbiting astronauts are not entirely weightless. When their space capsules are traveling around low-altitude orbits, subtle forces induced by atmospheric drag and solar radiation pressure create a total acceleration of about one-millionth of a g. Consequently, a 150-pound astronaut weighs about 1/6000th of a pound when coasting around in a low-altitude orbit.

Even before the start of the Civil War, weightlessness was being used routinely in the manufacture of useful products including the cannon grapeshot later seen careening across Civil War battlefields. Those small metal spheres were produced by releasing molten globules of metal inside special drop towers. If you release a molten globule of metal in a drop tower 250 feet tall, it will experience free-fall weightlessness

---

[1] Don't try this in your friendly, neighborhood hotel!

for about four seconds. At the bottom of the tower, it plunges into a vat of water where it quickly solidifies.

Microgravity (weightlessness) of longer duration can be achieved by flying special airplanes along free-fall parabolic trajectories. If such an airplane ascends to a seven-mile altitude and then falls back down again, it can provide weightlessness for about 20 seconds. This was the approach used in filming the realistic free-fall sequences in the Hollywood movie *Apollo 13*.

By flying a sounding rocket along a free-fall trajectory 250 miles high, we can achieve weightlessness for as much as five minutes. Thousands of sounding rockets have been launched in this manner, in part, to test the feasibility of manufacturing useful products in the weightlessness of space.

If we launch a satellite into any orbit around Earth, we can produce microgravity of equal or superior quality for days, weeks, or even longer durations. Experiments flown aboard manned space stations or unmanned satellites are providing us with ample opportunities for manufacturing a variety of valuable products made in space.

2 SATELLITE ORBITS

> If I have managed to see farther than others have, it is because I am a midget standing on the shoulders of giants.—*Isaac Newton*
> *in 1698 at age 56, twenty-nine years before his death*

Edmund Halley, who became world famous for discovering the periodic comet that bears his name, visited Isaac Newton one afternoon to ask him about the shape of planetary orbits. "If a planet is coasting within an inverse square gravitational field," Halley asked his old friend, "what shape would its orbit assume?"

"An ellipse," Newton replied without a moments hesitation. He then went on to explain that he had, some years earlier, proven a simple mathematical theorem nailing down the orbit's shape. That proof was now buried somewhere among his many papers, but after a brief search, he gave up any hope of being able to find it that day. However, he promised to send a copy of it to Halley if it ever turned up. Newton never managed to locate his earlier notes, so he proved the theorem again, this time in a more refined and polished fashion.

Isaac Newton was, by all accounts, the original absent-minded professor. When a young colleague came for an unexpected visit, Newton was occupied in the next room. As he waited, the young visitor became increasingly famished. So he ate the food that had been set out for Isaac Newton's lunch. Much later, when Newton came back into the room and encountered the empty plate, he hardly reacted at all. "I thought I had not taken my meal," he observed. "But I see that I have after all."

ISAAC NEWTON'S *VIS VIVA* EQUATION

Through a brilliant line of reasoning, Isaac Newton derived the so-called *vis viva* equation, defining the instantaneous velocity of a satellite as it coasts around an elliptical orbit.[1] The *vis viva* equation can be expressed as follows:

$$V = \mu^{1/2}\left(\frac{2}{r} - \frac{1}{a}\right)^{1/2}$$

In this equation, V denotes the instantaneous velocity of the satellite, μ represents the gravitational parameter of Earth, r is the radial distance between the center of Earth and the center of mass of the satellite, and a is the semi-major axis of its elliptical orbit.

The value of the semi-major axis is obtained by measuring the length of the ellipse and then dividing by 2. Notice that once a satellite is traveling around its elliptical orbit, the only variables in the *vis viva* equation are the radial distance r and the velocity V.

As it stands, the *vis viva* equation can be enormously helpful for making simple calculations in orbital mechanics, and it can be further simplified for those special cases in which the satellite is traveling along a circular orbit or a parabolic escape trajectory.

If a satellite is in a circular orbit, its semi-major axis a will equal its radius r. Consequently, as the derivation in the upper right-hand corner of Figure 2.1 indicates, the orbital velocity of a satellite in a circular orbit at any radius r from the center of Earth is given by

$$V_{\text{CIRC}} = \left(\frac{\mu}{r}\right)^{1/2}$$

Thus, we see that satellites in higher-altitude circular orbits travel slower than those in lower circular orbits.

If you are riding around Earth in a circular orbit aboard your trusty rocketship and you mash down on the accelerator until

[1] In Latin, *vis viva* means "living force."

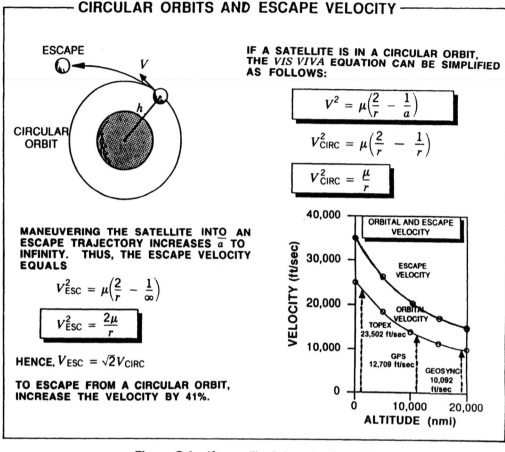

Figure 2.1 If a satellite is in a circular orbit its semi-major axis *a* will equal its radius *r*. Consequently, as the miniderivation at the top right shows, the orbital velocity of an Earth-orbiting satellite in a circular orbit of radius *r* always equals the square root of the ratio of Earth's gravitational parameter and the radius *r*. At that same radius *r*, the orbiting satellite will escape Earth if its velocity is increased by 41 percent.

your speed increases by 41 percent, you will escape Earth never to return. When your rocketship reaches the minimum velocity needed to escape from Earth, it will be traveling along an escape parabola, which is an ellipse with infinite length ($a = \infty$). Consequently, according the *vis viva* equation, the escape velocity at any radius *r* (see Figure 2.1) is given by

$$V_{ESC} = \left(\frac{2\mu}{r}\right)^{1/2}$$

By comparing the equation for a satellite's circular orbital velocity with the equation for its escape velocity, we can conclude that the escape velocity at any radius r is always equal to the $\sqrt{2}$ times the circular orbital velocity at that same radius. Consequently, a 41-percent increase in the velocity of a satellite traveling around a circular orbit at any altitude will always cause it to escape from Earth.

The small graph in the lower right-hand corner of Figure 2.1 depicts the circular orbital velocities and the escape velocities at various orbital altitudes. Notice that both velocity curves asymptotically approach a value of zero as the radius r approaches infinity. Notice also that the escape velocity curve is, at every radius, 1.41 (the square root of 2) times higher than the circular orbital-velocity curve.

EVALUATING THE GRAVITATIONAL PARAMETER OF EARTH

The value of the gravitational parameter (μ) in the *vis viva* equation is related in a simple manner to the radius of Earth, r_0, and its surface gravitational acceleration, g_0. Two different methods for evaluating this important constant are available.

As Isaac Newton noted, the inverse square law of gravitation can be written as

$$F_g = \frac{GMm}{r^2}$$

where F_g is the force of gravity pulling on a satellite of mass m at any radius r. In this equation, G is the universal gravitational constant (which applies everywhere in the universe), and M is the mass of Earth.

Newton's second law tells us that the gravitational acceleration experienced by a satellite is given by

$$g = \frac{F_g}{m} = \frac{GM}{r^2} = \frac{\mu}{r^2}$$

Consequently,

$$\mu = GM$$

The symbol μ, of course, denotes the gravitational parameter of the central body. In this case, it is the gravitational parameter of Earth.

Another way of evaluating the value of Earth's gravitational parameter is to note that for a small satellite in orbit about a large central body composed of shells of equal-density material, the gravitational acceleration at any radius r is given by

$$g = g_o \left(\frac{r_o^2}{r^2} \right)$$

Consequently, under these assumptions, the gravitational parameter must be given by

$$\mu = g_o r_o^2$$

If we assume that the gravitational acceleration of a body falling downward near the surface of Earth is 32.174 ft/sec² and Earth's radius is 20,925,741 feet, then the gravitational parameter for Earth is

$$\mu = g_o r_o^2 = \left(32.174 \ \frac{\text{ft}}{\text{sec}} \right) (20{,}925{,}741 \ \text{ft})$$

$$= 1.4076540 \times 10^{16} \ \text{ft}^3/\text{sec}^2$$

Every large, spherical body in the universe has its own gravitational parameter μ, which, as we have seen, is equal to $g_o r_o^2$, where g_o is the surface acceleration, and r_o is the radius of the spherical body.

Table 2.1 presents side-by-side comparisons between a tree-top satellite's velocity and the corresponding escape velocity for various spherical bodies in our solar system. Notice that the treetop velocity for a satellite in orbit about Earth (top line) is 25,936 feet per second. The escape velocity is $\sqrt{2}$ times that value—or 36,600 feet per second. The table also includes similar velocity values for the Moon, the Sun, Mercury, Venus, Mars, Jupiter, and an asteroid three miles in radius.

Never spend your summer vacation on the surface of Jupiter unless you have an extremely efficient rocket to help you get back home. The velocity required to escape from the surface of Jupiter is 200,000 feet per second. Escaping from the surface of the Sun is even more difficult. You will need a rocket that can generate 2 million feet per second to get your body back home again if your vacation destination lies on the surface of the Sun.

ESCAPING FROM THE SURFACE OF A THREE-MILE ASTEROID

The *vis viva* equation, together with some of the other relationships we have developed so far, can be used in calculating the velocity required to the escape from the surface of a three-mile asteroid. The two equations at the top of Figure 2.2 can be used in evaluating the local gravitational accelerations above its surface and on the inside, too.

If we assume that the asteroid in question has the same average density as Earth, we can use the gravitational acceleration equation on the left to make the necessary asteroid calculations. This is a valid approach because if you are three miles from the center of Earth, you will have the same gravitational acceleration acting on your body as you would if you were standing on the surface of a three-mile asteroid. Consequently, the gravitational acceleration you will experience at the surface of the asteroid is given by

$$g_A = g_0 (R_A/R_0) = 0.024 \text{ ft/sec}^2$$

Once we have computed this value for the local gravitational acceleration, we can determine the gravitational parameter of the asteroid (see Figure 2.2) from the following relationship:

$$\mu_A = g_A R_A^2 = 6.0217 \times 10^6 \text{ ft}^3/\text{sec}^2$$

TABLE 2.1 Approximate Gravitational Parameters and Escape Velocities for Various Heavenly Bodies

Heavenly Body	Acceleration of Gravity at Surface, g_o (ft/sec^2)	Equatorial Radius, r_o (miles)	$\mu = g_o r_o^2$ Approximate Gravitational Parameter (ft^3/sec^2)	$V_{c_o} = \left(\dfrac{\mu}{r_o}\right)^{1/2}$ Treetop Orbital Velocity (ft/sec)	$V_{e_o} = \left(\dfrac{2\mu}{r_o}\right)^{1/2}$ Escape Velocity at Surface (ft/sec)
Earth	32.174	3,963	1.4076540×10^{16}	25,936	36,600
Moon	5.19	1,080	1.69×10^{14}	5,440	7,800
Sun	897.1	432,500	4.68×10^{21}	1,434,750	2,023,000
Mercury	12.42	1,515	7.94×10^{14}	9,708	13,800
Venus	28.28	3,760	1.11×10^{16}	23,538	33,700
Mars	12.23	2,111	1.52×10^{15}	11,631	16,400
Jupiter	81.40	44,423	4.48×10^{18}	141,373	200,000
3 Mile Asteroid	0.024	3	6.02×10^{6}	19.5	27.5

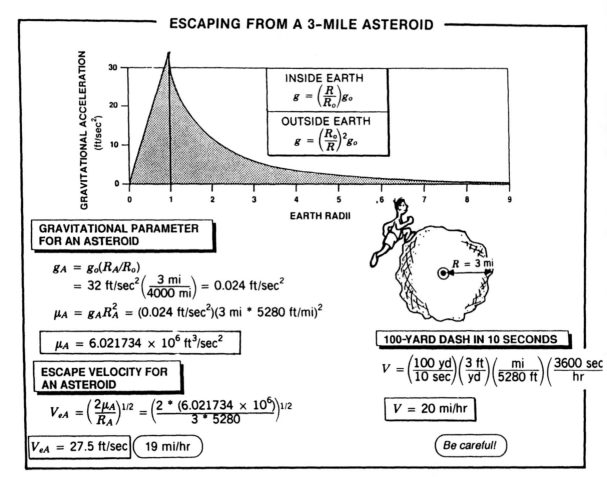

Figure 2.2 These calculations indicate that a runner who completes a 100-yard dash in 10 seconds could run off the surface of a small asteroid never to return. The relationships used in the derivation employ the inverse square law of gravitation, the equation defining the gravitational parameter of a spherical celestial body, and Isaac Newton's *vis viva* equation.

Notice that for an asteroid three miles in radius, the gravitational parameter is roughly 10 orders of magnitude smaller than the more familiar gravitational parameter for Earth.

The equation defining the velocity needed to escape from the surface of the asteroid is presented at the bottom of Figure 2.2. That equation can be evaluated as follows:

$$V_{e_A} = \left(\frac{2\mu}{R_A}\right) = 27.5 \text{ ft/sec}$$

Thus, we see that escaping from the surface of a 3-mile asteroid requires a velocity of only 27.5 feet per second, or about 19 miles per hour.

If you can run a 100-yard dash in 10 seconds, be especially careful if you are vacationing on a small asteroid. If you forget where you are, you may go out for an afternoon of recreational jogging, become overly enthusiastic, and fly off into space, never to return!

ELLIPTICAL ORBITS

Isaac Newton's *vis viva* equation can be rearranged so the three terms it contains represent three different kinds of specific energy as follows:

$$
\underset{2a}{\underset{\text{Energy}}{\underset{\text{Total}}{-\mu}} } = \underset{2}{\underset{\text{Energy}}{\underset{\text{Kinetic}}{V^2}}} - \underset{r}{\underset{\text{Energy}}{\underset{\text{Potential}}{\mu}}}
$$

The specific energy of an object is its energy per unit mass. For simplicity, we will assume that the satellite has a mass of one pound. Notice that the satellite's total specific energy, which appears on the left-hand side of the equal sign, is a function only of the semi-major axis of its orbit. Once a satellite is traveling around a particular elliptical orbit, its semi-major axis and, in turn, its total specific energy do not vary.

The term immediately to the right of the equal sign ($V^2/2$) represents the kinetic energy of the one-pound satellite. The kinetic energy is a function only of the satellite's instantaneous velocity which, of course, varies in accordance with Kepler's law of equal areas. The term on the far right is the specific potential energy. Notice that it is a function only of the satellite's radial distance r. That radial distance, of course, varies continuously as the satellite travels around its elliptical orbit.

Isaac Newton derived the *vis viva* equation by noting that the sum of a satellite's kinetic energy and its potential energy must always equal a constant. In his famous derivation, he adopted the convention that the potential energy at infinity would equal zero. Consequently, the potential energy for a satellite located at any altitude below infinity is always a *negative* quantity. It would, of course, seem more natural to assume a zero potential energy at the surface of Earth because that is where we spend our lives. However, by setting the potential energy at infinity equal to zero, Newton simplified the *vis via* equation with a corresponding loss in physical insight.

The three curves running across Figure 2.3 represent the kinetic energy, the total energy, and the potential energy, respectively, for a one-pound satellite at different points along a particular elliptical orbit in which the perigee radius equals two Earth radii and the apogee radius equals four Earth radii. The satellite begins its journey at the perigee of its orbit where its true anomaly θ equals zero. At perigee, its kinetic energy ($V^2/2$) is about 6.2 million foot-pounds.

As it moves upward away from perigee, the satellite's velocity decreases, thus creating a corresponding decrease in its kinetic energy. This gradual reduction in kinetic energy persists until it reaches apogee where its true anomaly equals 180°. When it passes apogee and begins to fall back down toward perigee again, its velocity gradually increases with a corresponding increase in its kinetic energy.

Notice that the potential-energy curve, which is represented by the gently contoured line running across the bottom of Figure 2.3, is the mirror image of the kinetic energy curve running across the top. The potential energy of the satellite has its smallest value at perigee. Its value then gradually increases as it moves up toward apogee. Then, when it passes apogee and begins to fall back down toward perigee, its potential energy begins to shrink.

The total energy of the satellite (the algebraic sum of its kinetic energy and its potential energy) is defined by the curve running across the middle of Figure 2.3. Notice that this curve

---

Figure 2.3 When the tetherball, far right, travels upward along its tilted orbit, it is constantly trading kinetic energy for potential energy. Then, when it falls back down again, it makes the reverse trade. A similar energy trade-off occurs when a satellite in an elliptical orbit coasts from perigee up to apogee and then back down again. Notice that the kinetic-energy curve and the potential-energy curve (top and bottom of the graph, respectively) are mirror images of one another. Notice also that the total energy which is represented by the curve running across the center of the graph, always remains constant regardless of the position of the satellite.

ELLIPTICAL-ORBIT ENERGY TRADES

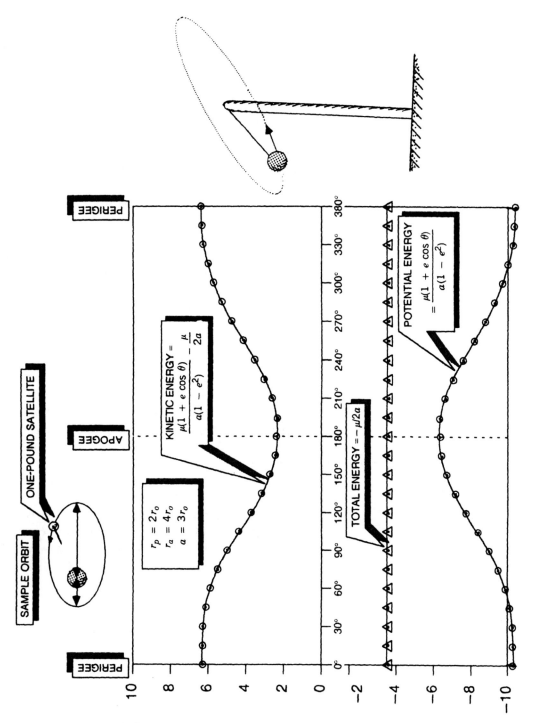

is a straight, horizontal line with a value of about -3.8 million foot-pounds. The total energy of an orbiting satellite, which is a function only of the magnitude of its semi-major axis, always remains constant once it enters an elliptical orbit with a particular semi-major axis.

THE HIGH COST OF LAUNCHING A SATELLITE

By carefully evaluating the *vis viva* equation, we can show that the energy required to launch a one-pound satellite from the surface of Earth into a low-altitude circular orbit equals approximately 10 million foot-pounds. Launching a similar one-pound mass from the surface of Earth onto an escape parabola requires 20 million foot-pounds per pound. Thus, we see that it takes twice as much energy to loft a payload from the surface of Earth onto an escape trajectory as it does to put that same payload into a low-altitude circular orbit.

Generating 10 million foot-pounds of energy may seem like an expensive proposition. So we might logically conclude that the cost of the orbital energy required constitutes a substantial fraction of the cost of launching a payload into space. But, actually, 10 million foot-pounds of energy can be surprisingly inexpensive. As Table 2.2 indicates, 10 million foot-pounds of ordinary household electricity amounts to only about 37 cents, if we assume the homeowner pays 10 cents per kilowatt-hour.

Heating oil, gasoline, and wood all cost somewhere between 13 and 21 cents for 10 million foot-pounds. The energy equivalent of 10 million foot-pounds of TNT costs about seven dollars. In view of these surprisingly low prices for 10 million foot-pounds of energy, why does it cost at least $3000 to place each pound of payload into a low-altitude orbit?

TABLE 2.2 How Much Does 10 Million Foot-Pounds of Energy Cost?*

Energy Source	Unit Cost	Amount Required for 10M ft-lb	Cost for 10M ft-lb
Home electricity	10 cents/kilowatt-hour	3.7 kilowatt-hours	37 cents
Heating oil	$1/gallon	1 pound	14 cents
Gasoline	$1/gallon	0.9 pound	13 cents
Wood	$300/cord	0.1 ft^3	21 cents
TNT	$1/pound	7 pounds	700 cents

*Putting a pound of payload into orbit requires 10 million ft-lb of energy. Typically, this costs about $3000.

The high cost of transporting payloads into space stems from a number of factors other than the cost of the fuel needed to generate the minimum amount of energy required. For one thing, when we launch a payload into space, the rocket must carry oxidizer in addition to the fuel. Moreover, the oxidizer required invariably outweighs the fuel by at least a factor of 2.

When you burn a pound of gasoline in your car, you also consume the oxygen contained in 21 pounds of air. The air, of course, comes free. But, when we burn the fuel in a rocket—kerosene or liquid hydrogen, for instance—we must also carry an appropriate amount of liquid oxygen, nitrogen tetroxide, or some other appropriate oxidizer.

A rocket fueled with kerosene, for example, must carry about 2.3 pounds of oxidizer (liquid oxygen) for every pound of fuel (liquid kerosene). A hydrogen-oxygen rocket burns about five pounds of liquid oxygen for every pound of liquid hydrogen. We must, of course, pay for and accommodate both fluids when we fly a rocket into space.

Another reason it costs so much to launch a satellite into orbit is that we must accelerate surprisingly large quantities of propellant (fuel and oxidizer) up to high velocities so they can be later burned by the rocket. Approximately 97 percent of the energy generated by a typical booster is wasted in this largely unproductive manner.

Costs further escalate because we must pay for expensive engines, tanks, pumps, computers, guidance systems, and so on, which, for most rockets, are destroyed on each flight. Reusable space shuttles reduce the amount of lost hardware, but, compared with an expendable rocket, a reusable space shuttle typically carries less than half as much useful payload per pound of liftoff weight.

A modern multistage expendable booster is usually able to place less than three percent of liftoff weight into a low-altitude orbit. By contrast, reusable space shuttles such as the ones designed and operated by the Soviet Union and the United States can, more typically, orbit only about 1.2 percent of their liftoff weight.

Other important cost factors also come into play. Operating a large, complicated rocket safely and reliably requires a standing army of engineers and technicians who are on the payroll even when satellites are not being launched. Moreover, launch pads, tracking stations, security services, and mission control centers must be operated and maintained regardless of current launch rates.

THE BEST PLACE TO ESCAPE FROM AN ELLIPTICAL ORBIT

Isaac Newton's *vis viva* equation provides us with a convenient tool for determining the best locations and the best techniques for performing powered flight maneuvers in space.

Assume for the moment that you are an astronaut whirling around Earth in the specific elliptical orbit shown in Figure 2.4. As you can see, its perigee radius equals twice the radius of Earth and its apogee radius equals four times Earth radius. Under the stated conditions, where is the best place to fire your rocket to drive your spaceship onto an infinitely elongated escape parabola?

1. At apogee?
2. At perigee?
3. Somewhere in between?
4. In the sunlight so you can see what you are doing?
5. It doesn't matter where the maneuver is executed, because escaping from an orbit always takes the same amount of energy?

The calculations in Figure 2.4 compare the impulsive ΔV's required for an escape from apogee and an escape from perigee. Notice that each of these two calculations uses the *vis viva* equation two different times.

In the first calculation, which is summarized at the top of Figure 2.4, the velocity increment we need to add at apogee equals the escape velocity we need at that orbital altitude minus the apogee velocity we already have. Escaping from apogee requires an additional velocity increment of about 7750 feet per second. Escaping from perigee (see Figure 2.4) requires a velocity increment of only 4759 feet per second. Thus, we see that, in this case, a clever rocketeer can save approximately 3000 feet per second by escaping at perigee rather than at apogee. Escaping from the perigee of an elliptical orbit, it turns out, always requires a smaller velocity increment.

In the early days of the U.S. space program when my friend Bob Africano and I were struggling to master the intricacies of orbital mechanics, we developed a simple rule of thumb for executing spaceborne maneuvers more efficiently:

Whenever possible, always trade your potential energy for kinetic energy, then add the rocket ΔV.

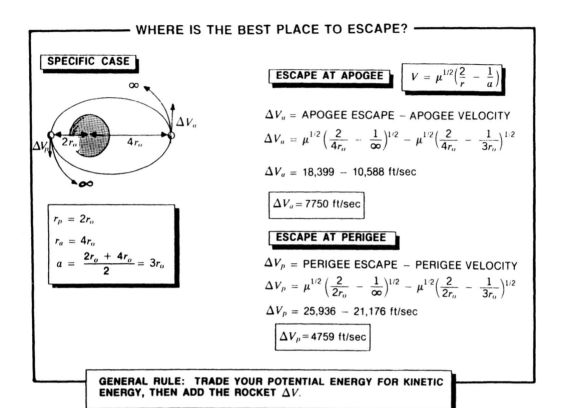

WHERE IS THE BEST PLACE TO ESCAPE?

SPECIFIC CASE

$r_p = 2r_o$

$r_a = 4r_o$

$a = \dfrac{2r_o + 4r_o}{2} = 3r_o$

ESCAPE AT APOGEE $V = \mu^{1/2}\left(\dfrac{2}{r} - \dfrac{1}{a}\right)$

ΔV_a = APOGEE ESCAPE – APOGEE VELOCITY

$\Delta V_a = \mu^{1/2}\left(\dfrac{2}{4r_o} - \dfrac{1}{\infty}\right)^{1/2} - \mu^{1/2}\left(\dfrac{2}{4r_o} - \dfrac{1}{3r_o}\right)^{1/2}$

ΔV_a = 18,399 – 10,588 ft/sec

$\Delta V_a = 7750$ ft/sec

ESCAPE AT PERIGEE

ΔV_p = PERIGEE ESCAPE – PERIGEE VELOCITY

$\Delta V_p = \mu^{1/2}\left(\dfrac{2}{2r_o} - \dfrac{1}{\infty}\right)^{1/2} - \mu^{1/2}\left(\dfrac{2}{2r_o} - \dfrac{1}{3r_o}\right)^{1/2}$

ΔV_p = 25,936 – 21,176 ft/sec

$\Delta V_p = 4759$ ft/sec

GENERAL RULE: TRADE YOUR POTENTIAL ENERGY FOR KINETIC ENERGY, THEN ADD THE ROCKET ΔV.

Figure 2.4 An astronaut coasting around an elliptical orbit can always reach escape velocity most efficiently if she makes a single impulsive burn near the perigee of her orbit. In this case, the astronaut occupies an orbit with a perigee radius is equal to $2r_o$ and an apogee radius equal to $4r_o$. As these calculations indicate, the astronaut escaping at perigee would save approximately 3000 feet per second compared with her poorly informed colleague who executes a similar impulsive maneuver at apogee hoping to escape.

This general rule of thumb can sometimes produce dramatic increases in the performance capabilities of a rocket.

Ambitious rocketeers have perfected a clever experiment to verify that we can get more performance from a rocket if we trade its potential energy for kinetic energy, then add the rocket ΔV.

They accomplish this by dropping two basketballs positioned one atop the other from a balcony located high above a basketball court. The ball on the bottom hardly bounces at all. But the one on top is hurled upward at impact at an unexpectedly high altitude. In this experiment, the potential energy of both balls is converted into kinetic energy, virtually all of which ends up in the top ball.

GRAVITY WELLS

Gravity wells, which are frequently found on display in aerospace museums, provide us with a concrete means for understanding the laws governing the movements of unpowered satellites and the effects of powered flight maneuvers in space. A gravity well (see Figure 2.5) is a whirlpool-shaped contour whose curving sides have a very specific mathematical shapes. Marbles rolling around the walls of the gravity well behave much like orbiting satellites. Museum visitors are often oblivious to what concept a gravity well is meant to teach, but, nevertheless, they find fascination in the curving trajectories of the balls rolling along its contoured surface.

Years ago, I suggested that the producers at ABC TV construct a large, transparent gravity well, so large that one of their

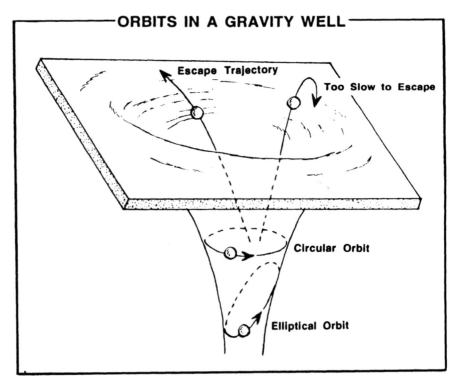

ORBITS IN A GRAVITY WELL

Escape Trajectory

Too Slow to Escape

Circular Orbit

Elliptical Orbit

Figure 2.5 A gravity well is a whirlpool-shape contour used to simulate the inverse square gravitational field of Earth using the constant gravitational field we all occupy at its surface. Marbles thumped along the inner walls of the gravity well behave in a manner that mimics the behavior of satellites coasting along circular, elliptical, escape, or sounding-rocket trajectories.

talk-show hosts, Ed Nelson, could go inside and thump marbles at different velocities to stimulate various types of maneuvers being executed by the astronauts. If he placed a marble on the walls of the gravity well and thumped it with exactly the right horizontal velocity, for example, it would go into a circular orbit, as shown in the middle of Figure 2.5.

If he thumped it a bit harder, it would swing in an elliptical orbit, first climbing up the walls of the gravity well and then rolling back down again. If he thumped the marble straight up along the walls of the gravity well at an intermediate speed, it would coast up and then fall back down again. That is a reasonably accurate simulation of the way a sounding rocket behaves. If he thumped the marble a bit harder, however, it would reach escape velocity. In this case, it will climb forever up the walls of the gravity well. As it climbs, it will always be slowing down, but it will never slow down to zero.

Figure 2.6 shows how much energy it takes travel from various altitudes in the vicinity of Earth. Earth's gravity well is 4000 miles (20 million feet) deep. That particular depth arises because it requires approximately 20 million foot-pounds per pound to drive a payload onto an escape trajectory.

In terms of energy, the international space station is half way to infinity. It lies half way up along the walls of Earth's gravity well. Consequently, launching a one-pound mass from the surface of Earth to the space station requires an energy expenditure of 10 million foot-pounds.

In terms of energy, the geosynchronous altitude (19,300 nautical miles) is positioned essentially at infinity. Thus, it takes approximately the same amount of energy to drive a one-pound mass up to the geosynchronous altitude as it does to drive that same one-pound mass onto an escape parabola.

As Figure 2.6 indicates, the gravity well of the Moon is about 1/20th as deep as the gravity well of Earth. Thus, it takes only 1/20th as much energy to hurl any mass off the surface of the Moon as it does to hurl that same mass off the surface of Earth. Surprisingly, it requires less energy to launch a payload off the Moon, fly it to the geosynchronous altitude, and circularize its orbit, than it does to fly that same payload off the surface of Earth into a geosynchronous orbit.

KEPLER'S EQUATION

In devising his three simple laws of planetary motion, Johannes Kepler worked with three different angles that are still of cru-

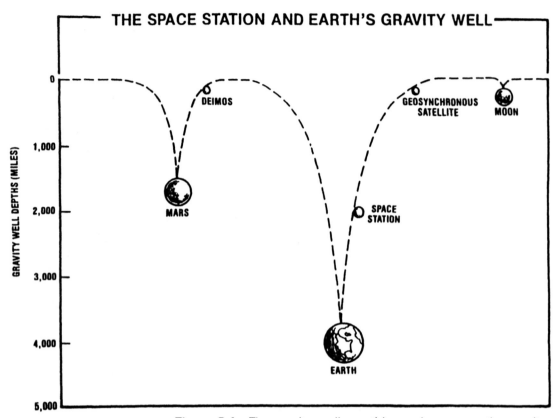

Figure 2.6 The gravity wells on this graph compare the gravitational fields of Mars, Earth, and the Moon. Note that the Moon's gravity well is only one-twentieth as deep as the gravity well of Earth. This implies that escaping from the surface of the Moon requires only one-twentieth as much energy as launching a comparable payload up to escape velocity from the surface of Earth. The gravity well for Mars is approximately 45 percent as deep as the gravity well of Earth.

Figure 2.7 These three angles—the true anomaly, the eccentric anomaly, and the mean anomaly—are widely used in making orbital mechanics calculations. The true anomaly is the most familiar angle. It is the Earth-centered angle measured to the position of the satellite from the line of apsides pointing toward perigee. The eccentric anomaly is constructed by circumscribing a circle around the elliptical orbit and then dropping a vertical line from the circle through the satellite to the semi-major axis line. A second line is then drawn from the intersection point on the circle to the center of the ellipse. The mean anomaly, a fictitious angle, is always proportional to that fraction of the orbital period that has elapsed since the satellite last passed the perigee of its orbit.

THREE CRITICAL ANGLES IN ORBITAL MECHANICS

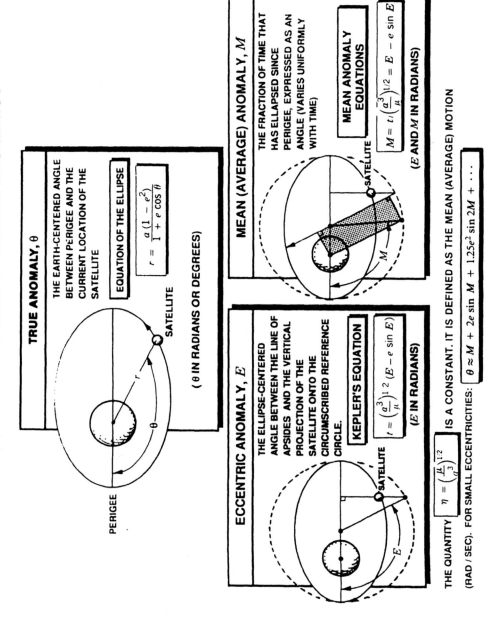

TRUE ANOMALY, θ

THE EARTH-CENTERED ANGLE BETWEEN PERIGEE AND THE CURRENT LOCATION OF THE SATELLITE

EQUATION OF THE ELLIPSE

$$r = \frac{a(1 - e^2)}{1 + e \cos \theta}$$

(θ IN RADIANS OR DEGREES)

PERIGEE

SATELLITE

r

θ

ECCENTRIC ANOMALY, E

THE ELLIPSE-CENTERED ANGLE BETWEEN THE LINE OF APSIDES AND THE VERTICAL PROJECTION OF THE SATELLITE ONTO THE CIRCUMSCRIBED REFERENCE CIRCLE.

KEPLER'S EQUATION

$$t = \left(\frac{a^3}{\mu}\right)^{1/2}(E - e \sin E)$$

(E IN RADIANS)

SATELLITE

E

MEAN (AVERAGE) ANOMALY, M

THE FRACTION OF TIME THAT HAS ELLAPSED SINCE PERIGEE, EXPRESSED AS AN ANGLE (VARIES UNIFORMLY WITH TIME)

MEAN ANOMALY EQUATIONS

$$M = t\left(\frac{a^3}{\mu}\right)^{1/2} = E - e \sin E$$

(E AND M IN RADIANS)

SATELLITE

M

THE QUANTITY $\eta = \left(\frac{\mu}{a^3}\right)^{1/2}$ IS A CONSTANT. IT IS DEFINED AS THE MEAN (AVERAGE) MOTION

(RAD / SEC). FOR SMALL ECCENTRICITIES: $\theta \approx M + 2e \sin M + 1.25e^2 \sin 2M + \cdots$

47

cial importance to modern practitioners of orbital mechanics. He also developed a special transcendental equation, which we have named Kepler's equation in his honor. Kepler's equation provides us with powerful insights into the behavior of solid bodies moving through space. In addition, over the centuries, it has sparked a rich heritage of mathematical research.

In order to understand Kepler's equation, it is necessary to study and appreciate the three angles shown in Figure 2.7: the *true anomaly*, the *eccentric anomaly*, and the *mean anomaly*.

The use of the word *anomaly* to denote an angle came into use during Kepler's lifetime, but it is not clear whether Kepler himself invented that particular use of the word. As Figure 2.7 indicates, all three of the angles in question are measured from the perigee of the elliptical orbit. The true anomaly (θ) is the easiest angle to understand. The equation linking the satellite's radial distance to its current true anomaly is merely the polar coordinate equation of an ellipse:

$$r = \frac{a(1 - e^2)}{1 + e \cos \theta}$$

The eccentric anomaly, E, is also depicted in Figure 2.7. "Eccentric" means *weird* and "anomaly" means *weird*, so the "eccentric anomaly" is *weird weird*! We begin the construction of the eccentric anomaly by circumscribing a circle around the elliptical orbit. As Figure 2.7 indicates, this circle is tangent to the two ends of the ellipse—at perigee and at apogee. Once this circle is in place, we erect a perpendicular from the line of apsides through the satellite to intersect the circle at a point immediately above (or below) the satellite. Finally, we draw a straight line through this intersection point to the center of the ellipse.

When this rather complicated construction has been completed, we measure the eccentric anomaly, E, from the perigee point in the direction of travel of the satellite to the line joining the center of the ellipse to the intersection point. The equation linking the eccentric anomaly and the travel time, t (see Figure 2.7), is called *Kepler's equation*. This equation is discussed further in a later section.

The third important angle, the mean anomaly, can be described as follows. First, we establish the orbital period of the satellite. Then we build a special "clock" with only one hand that travels through one complete circle at a steady rotation rate so that it completes one 360° revolution in exactly one orbital period.

If the orbital period of the satellite is 90 minutes, for instance, we would build a clock with a hand that travels around a complete 360° circle in 90 minutes. At any time, the mean anomaly is the number of degrees through which that hand has turned at its constant rotational rate since the satellite passed perigee.

Figure 2.8 highlights the varying rates at which these three different angles increase for a particular satellite orbit—the Soviet Molniya orbit, which is routinely used by the Russians for their constellations of communication satellites. A satellite in a Molniya orbit travels around Earth over an interval of 12 hours. Typically, the perigee altitude of such a satellite is 500 nautical miles and its apogee altitude is 21,360 nautical miles.

Notice that the horizontal scale in Figure 2.8 spans a period of 12 hours, whereas the vertical scale ranges from 0° to 360°. Three different angles associated with the Molniya orbit are plotted in Figure 2.8: the mean anomaly, the true anomaly, and the eccentric anomaly.

Not surprisingly, the mean anomaly curve is a straight diagonal line. This is true because the mean anomaly always evolves linearly with time. Notice that the true anomaly exhibits the most extreme excursions from the mean anomaly line, and that the eccentric anomaly more or less splits the difference between the other two curves. All three of the curves in Figure 2.8 have the same values when the satellite is at perigee (0 and 12 hours) and apogee (6 hours). The maximum excursions occur half way in between these other three time points at three and at nine hours, respectively.

Johannes Kepler used several mathematical theorems that had been proven by the ancient Greeks when he developed his closed-form relationship defining the location of a satellite traveling around an elliptical orbit. That relationship, which we call Kepler's equation in his honor, can be expressed as follows:

$$t = \left(\frac{a^3}{\mu}\right)^{1/2} (E - e \sin E)$$

When we are evaluating this equation, we normally know the time, t, but not the eccentric anomaly, E—which is embedded in two places on the right-hand side of Kepler's equation. For-

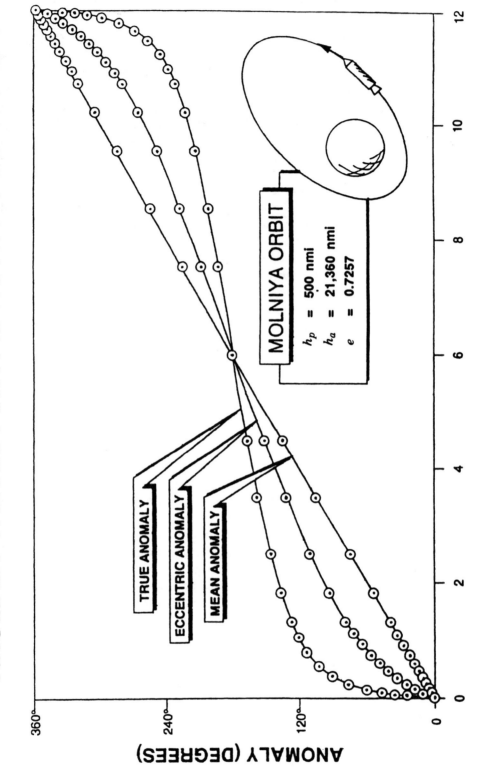

ANOMALY VALUES FOR A 12-HOUR MOLNIYA ORBIT

MOLNIYA ORBIT

h_p = 500 nmi
h_a = 21,360 nmi
e = 0.7257

TRUE ANOMALY

ECCENTRIC ANOMALY

MEAN ANOMALY

ANOMALY (DEGREES)

TIME SINCE PERIGEE (HOURS)

50

tunately, the equation can be solved iteratively or by series expansion.

Once we have used a series expansion or an iterative technique to solve Kepler's equation for E, we can compute the corresponding radial distance and the true anomaly using these two simple equations:

$$r = a(1 - e \sin E)$$

$$\tan \frac{\theta}{2} = \left(\frac{1 + e}{1 - e} \right)^{1/2} \tan \frac{E}{2}$$

The right-hand side of Figure 2.9 provides us with a convenient pictorial representation of Kepler's law of equal areas. The elliptical orbit in the figure is sliced into 18 short arcs, each representing 20 minutes of travel time for the satellite as it moves around Earth. Each of these 20-minute travel arcs is attached to a small wedge with its apex at the center of Earth. All 18 wedges have exactly the same area.

At first glance, Kepler's equation:

$$t = \left(\frac{a^3}{\mu} \right)^{1/2} (E - e \sin E)$$

seems deceptively simple. However, solving it is more complicated and tedious than you might think, because the variable we are seeking is the eccentric anomaly, E, which is embedded in two places on the right-hand side of Kepler's equation. It is impossible to manipulate Kepler's equation so that the eccentric anomaly, E, will appear on the left-hand side of the equal sign all by itself.

Isaac Newton derived one widely used method of iteration that is today called *Newton's method*. It works extremely well in many simple situations. In his book, *An Introduction to the*

Figure 2.8 This graph presents the true anomaly, the eccentric anomaly, and the mean anomaly over a single 12-hour orbit for a Soviet Molniya satellite with a low-altitude perigee and an apogee slightly above the geosynchronous altitude. Notice that the mean anomaly is represented by a straight, diagonal line. The true anomaly has the greatest curvature, whereas the eccentric anomaly essentially splits the difference between the other two curves.

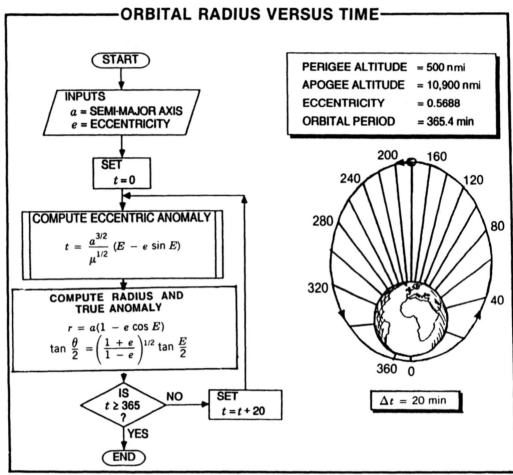

Figure 2.9 This simple computational algorithm shows how a series of points can be positioned 20 minutes apart along an elliptical orbit using Kepler's equation repeatedly to obtain the proper eccentric anomaly. In Kepler's equation, the time, t, is known, but the eccentric anomaly, E, is unknown. The double-walled flowchart box indicates that a subroutine must be called to obtain the correct value for E. That subroutine usually contains either a series expansion or an iterative procedure to solve for E.

Mathematics and Methods of Astrodynamics, Richard H. Battin provides us with a series expansion to allow us to solve for the value of the eccentric anomaly

$$E = M + \frac{e \sin M}{1 - e \cos M} - \frac{1}{2}\left(\frac{e \sin M}{1 - e \cos M}\right)^3 + \cdots$$

Generally speaking, the more terms we retrain in this series, the more accuracy we can achieve in solving for the eccentric anomaly. However, in some practical situations, the point of diminishing returns is quickly reached.

THE SIX CLASSICAL KEPLERIAN ORBITAL ELEMENTS

Using his newly perfected integral calculus, Isaac Newton showed that a small satellite in orbit about a large spherical body will trace out the same elliptical orbit over and over again. It may seem strange that only six numbers are required to completely characterize something as complicated as the future movements of a satellite. However, knowing those six orbital elements is tantamount to knowing its initial position coordinates, x, y, and z, and its initial velocity components $\dot{x}, \dot{y}$, and $\dot{z}$ together with the only force acting on the satellite—which, in this case, is the central body's gravitational attraction.

The six Keplerian orbital elements (ephermeris constants) are defined in Figure 2.10. The *semi-major axis, a*, is the half of length of the ellipse. Measure the elliptical orbit from one end to the other, and then divide by 2. The result is the semi-major axis of the satellite's orbit.

The *eccentricity, e*, defines the shape of the orbit. The orbital eccentricity equals the apogee radius, r_a, minus the perigee radius, r_p, divided by their sum. The eccentricity for a circular orbit is always equal to 0. The eccentricity for a parabolic escape orbit (infinitely elongated ellipse) is always equal to 1. All elliptical orbits, therefore, have eccentricities somewhere between 0 and 1.

The *inclination, i*, of a satellite's orbit is defined as the angle between its orbit plane and Earth's equatorial plane. For posigrade orbits (those in which the satellite generally travels in the same direction as Earth's rotation), the orbital inclination angle lies between 0° and 90°. For retrograde orbits (those in which the satellite travels generally in the opposite direction), the orbital inclination lies between 90° and 180°.

The *right ascension of the ascending node* (Ω) is the angle measured in the equatorial plane between the vernal equinox and the point at which the satellite orbit crosses the equator moving from the Southern Hemisphere into the Northern Hemisphere. The vernal equinox is the point at which Earth crosses the Earth–Sun (ecliptic) plane on the first day of spring.

The *argument of perigee* (ω) is measured in the orbit plane from the ascending node to the current perigee of the orbit.

THE SIX KEPLERIAN ORBITAL ELEMENTS

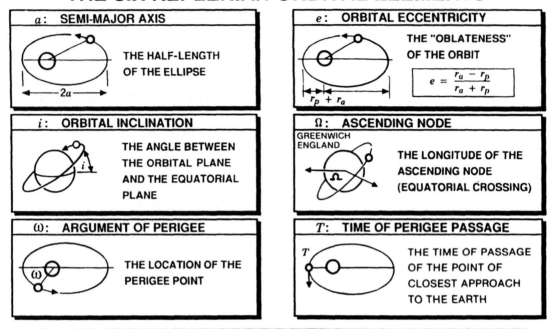

a : SEMI-MAJOR AXIS THE HALF-LENGTH OF THE ELLIPSE 2a	**e : ORBITAL ECCENTRICITY** THE "OBLATENESS" OF THE ORBIT $e = \dfrac{r_a - r_p}{r_a + r_p}$ $r_p + r_a$
i : ORBITAL INCLINATION THE ANGLE BETWEEN THE ORBITAL PLANE AND THE EQUATORIAL PLANE	**Ω : ASCENDING NODE** GREENWICH ENGLAND THE LONGITUDE OF THE ASCENDING NODE (EQUATORIAL CROSSING)
ω : ARGUMENT OF PERIGEE THE LOCATION OF THE PERIGEE POINT	**T : TIME OF PERIGEE PASSAGE** THE TIME OF PASSAGE OF THE POINT OF CLOSEST APPROACH TO THE EARTH

- **THESE SIX ORBITAL ELEMENTS ARE MATHEMATICALLY EQUIVALENT TO 3-D POSITION AND VELOCITY** $(x, y, z, \dot{x}, \dot{y}, \dot{z})$
- **THE GPS EPHEMERIS COORDINATES INCLUDE 4 EXTRA ELEMENTS FOR CLOCK ERRORS AND ABOUT 10 OTHERS FOR ORBITAL PERTURBATIONS**

Figure 2.10 Only six orbital elements are required to pin down the motion of a satellite traveling around a simple, two-body elliptical orbit. The shape and the size of the satellite's orbit are defined by its semi-major axis and its orbital eccentricity, e. The spatial orientation of the orbit is defined by the orbit's inclination, the ascending node, and the argument of perigee. Finally, the time of perigee passage defines the time at which the satellite is at its point of closest approach to Earth.

Finally, the symbol *T* denotes the *time of perigee passage* or the time at which the satellite reaches its point of closest approach to the center of Earth.

These six classical Keplerian orbital elements: *a, e, i,* Ω, ω, and *T* can be used in predicting the movement of a satellite around a simple two-body orbit. These six constants also can be used in calculating the satellite's three instantaneous position coordinates (*x, y,* and *z*) and its three instantaneous velocity components ($\dot{x}, \dot{y},$ and $\dot{z}$).

The flowchart in Figure 2.11 spotlights the computational algorithm that can be used in calculating the position and the velocity of the satellite at any time, t. Notice that the inputs are the six classical Keplerian orbital elements. In this algorithm, the true anomaly, θ, has been initialized at a value of $0°$. Given this initial value, the radial distance, r, from the center of Earth is calculated using the polar-coordinate equation for the elliptical orbit of the satellite.

In the next two flowchart boxes, the three position coordinates (x, y, and z) and the three corresponding velocity components ($\dot{x}$, $\dot{y}$, and $\dot{z}$) are calculated using a set of simple trigonometric equations. These equations can be derived using coordinate rotation matrices. Once these values have been printed, a test is made to see if the current true anomaly (θ) is

COMPUTING THE POSITION AND VELOCITY OF A SATELLITE

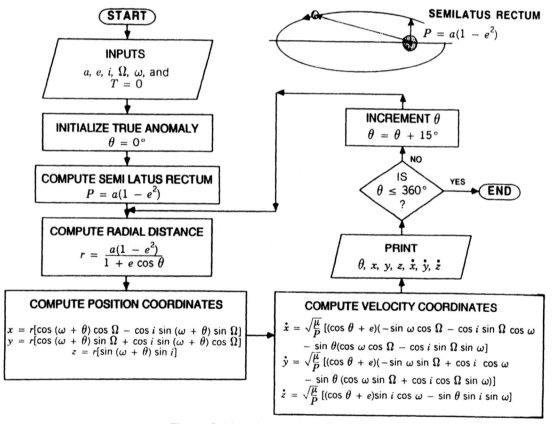

Figure 2.11 This simple computational algorithm shows how the six classical Keplerian orbital elements can be used to compute the three mutually orthogonal position coordinates and the three mutually orthogonal velocity components of a satellite corresponding to any true anomaly value defining the satellite's position along its orbit.

greater than or equal to 360°. If so, the computations halt. If not, θ is incremented by 15°, and the complete cycle of computations is repeated.

LAUNCH AZIMUTHS AND GROUND-TRACE GEOMETRY

All other things being equal, it always takes less ΔV to launch a satellite into orbit in the due eastward direction from a launch pad situated anywhere on Earth. This is true because Earth's rotation helps sling the satellite into orbit. For a launch pad on the equator, this free velocity increment amounts to 1520 feet per second, or about 1000 miles per hour!

If the launch pad is at some other latitude above or below the equator, the free velocity that can be obtained by launching in the eastward direction amounts to 1520 feet per second multiplied by the cosine of the latitude of the launch pad. Specifically, if a launch pad is located at a latitude of 30° (north or south of the equator), the free velocity that can be obtained turns out to be 1316 feet per second.

When we launch in the due east direction, our launch azimuth equals 90°. The launch azimuth is the angle between the boost trajectory plane and a vector pointing from the launch pad toward the North Pole. If we launch a satellite along any launch azimuth other than 90°, it will enter an orbit that has an inclination angle greater than the latitude of the launch pad.

When a satellite is launched into a low-altitude orbit, it will typically travel around Earth in about 90 minutes. Seen from inertial space, such a satellite will trace out essentially the same elliptical orbit over and over again. However, its ground trace will appear to drift in the westward direction. This gradual westward drift arises because, as the satellite coasts around its orbit, Earth rotates out from under its orbit plane.

On a Mercator projection, the ground trace for a low-altitude circular orbit resembles a shifting sine curve. During each orbit, which lasts approximately 90 minutes, the ground trace will be displaced in the westward direction by about 22.5°.

The parametric graphs in Figure 2.12 depict the ΔV increments that are provided free by Earth's rotation when satellites are launched into low-altitude orbits using various launch techniques. The curving line running across the top of the figure represents due-east launches from launch pads at various latitudes.

ΔV ASSIST FROM EARTH'S ROTATIONAL VELOCITY

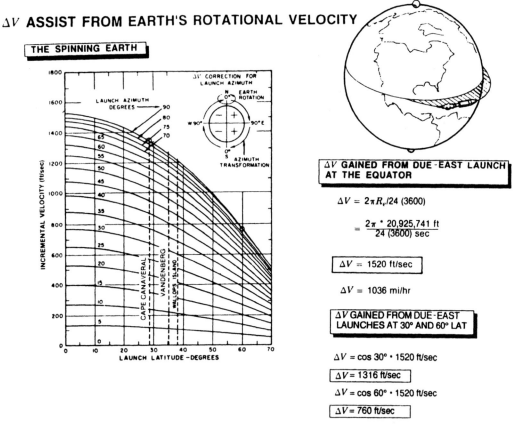

THE SPINNING EARTH

ΔV CORRECTION FOR LAUNCH AZIMUTH

AZIMUTH TRANSFORMATION

ΔV GAINED FROM DUE-EAST LAUNCH AT THE EQUATOR

$$\Delta V = 2\pi R_e/24 \,(3600)$$

$$= \frac{2\pi * 20{,}925{,}741 \text{ ft}}{24\,(3600) \text{ sec}}$$

$$\boxed{\Delta V = 1520 \text{ ft/sec}}$$

$$\Delta V = 1036 \text{ mi/hr}$$

ΔV GAINED FROM DUE-EAST LAUNCHES AT 30° AND 60° LAT

$$\Delta V = \cos 30° * 1520 \text{ ft/sec}$$

$$\boxed{\Delta V = 1316 \text{ ft/sec}}$$

$$\Delta V = \cos 60° * 1520 \text{ ft/sec}$$

$$\boxed{\Delta V = 760 \text{ ft/sec}}$$

Figure 2.12 The best place to build a launch pad for hurling satellites into orbit is on the equator, and the best launch direction is always due east. This is true because a due-east launch from the equator provides the maximum rotational velocity to help sling the satellite into orbit. The parametric graph on the right-hand side of this figure defines the free-velocity increments that can be obtained for satellites lofted along different launch azimuths from launch pads constructed at different latitudes.

Notice that, as was previously computed, a due-east launch from the equator provides a free velocity increment of 1520 feet per second. Similarly, a due-east launch from a pad at 30° north or south latitude provides a free-velocity increment of 1316 feet per second. Satellites launched due east from a 60° latitude gain only about 760 feet per second. The other contour lines in Figure 2.12 correspond to launches along azimuths other than 90°.

The allowable launch azimuths from a particular launch site are usually constrained to preclude crashing stray boosters into populated regions. Boosters launched from Cape Canaveral, for instance, can usually travel along azimuths ranging from 35° to 120°. The Western Test Range at Vandenburg Air Force Base in California cannot take advantage of Earth's rotation because populated regions lie east of the Vandenburg launch pads. However, Vandenburg can be used for launching satellites into near polar and sun synchronous orbits.

ORBITAL PERTURBATIONS

If we launch a satellite into orbit around a spherical planet with no forces of perturbation, the six classical Keplerian orbital elements completely characterize its orbit. In the real world, of course, perturbations do exist. Common examples include atmospheric drag, solar radiation pressure, Earth's equatorial bulge, and the higher-order potential harmonics (shape terms) of Earth. The perturbations induced by the gravitational attractions of the Sun and the Moon also must be taken into account when mission specialists are planning and executing some demanding space flight missions.

Three practical methods are available for modeling the effects of these and various other perturbations:

1. Frequently updating the six classical Keplerian orbital elements
2. Using an expanded set of the six orbital elements
3. Relying on numerical integration techniques to propagate the motion of the satellite

The engineers who masterminded the Navstar Global Positioning System (GPS) chose to employ the first two techniques. The orbital elements of each Navstar satellite are updated once per day from Falcon Air Force Base in Colorado. When the technicians there update a satellite, they provide it with 16 orbital elements, defining its current orbit to an accuracy of about 15 feet. The Russian engineers who designed the 24-satellite Glonass constellation decided, instead, to use numerical integration to propagate the orbital motion of their Glonass satellites.

The largest perturbation acting on the satellites in the Navstar constellation is the second zonal harmonic—the bulge at Earth's equator. As the sketch on the left-hand side of Figure 2.13 indicates, if we launch a satellite into a low-altitude orbit above a spherical Earth, it will trace out the same elliptical orbit

NODAL REGRESSION DUE TO EARTH'S EQUATORIAL BULGE

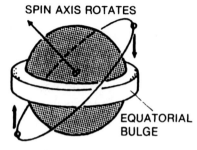

SPIN AXIS ROTATES

EQUATORIAL BULGE

FOR THE GPS:

$$\dot{\Omega} = \frac{-2.38247 \times 10^{13}}{a^{3.5}} \cos i$$

$$= \frac{-2.38247 \times 10^{13}}{(3440 + 10898)^{3.5}} \cos 55°$$

$$= 0.03872°/\text{day}$$

$$\boxed{\dot{\Omega} = 14°/\text{year}}$$

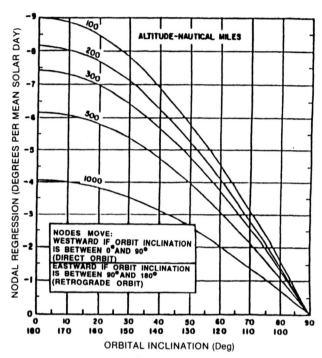

NODAL REGRESSION (DEGREES PER MEAN SOLAR DAY)

ALTITUDE-NAUTICAL MILES

NODES MOVE:
WESTWARD IF ORBIT INCLINATION IS BETWEEN 0° AND 90° (DIRECT ORBIT)
EASTWARD IF ORBIT INCLINATION IS BETWEEN 90° AND 180° (RETROGRADE ORBIT)

ORBITAL INCLINATION (Deg)

Figure 2.13 To a first approximation, a satellite launched into an elliptical orbit around a spherical planet will trace out repeatedly the same elliptical orbit fixed in inertial space. However, if the planet has an equatorial bulge, the satellite will be pulled down toward the bulge when it is over the Northern Hemisphere and it will be pulled up toward the bulge when it is over the Southern Hemisphere. This will cause the nodal crossing point to march systematically around the equator. A satellite in a 30° inclined orbit, at an altitude of 500 nautical miles, for example, will experience a nodal regression rate equal to 5.5° per mean solar day.

repeatedly provided we assume that there are no forces of perturbation to disturb its simple two-body motion.

However, if we launch that same satellite into orbit around an Earth with an equatorial bulge, it will experience perturbations that cause its orbit plane to twist systematically in space. As Figure 2.13 indicates, when the satellite is over the Northern Hemisphere, it is pulled *down* toward the equatorial bulge. When its over the Southern Hemisphere, it is pulled *up* toward the bulge. The net result is a gyroscopic force of precession that causes the nodal crossing point to march systematically around the equator. Another way of visualizing this phenomenon is to note that the spin-axis vector at the center of Earth, which is perpendicular to the satellite's orbit plane, traces out the same cone over and over again.

The nodal regression induced by Earth's equatorial bulge is not necessarily a small effect. As the graph on the right-hand side of Figure 2.12 indicates, a satellite in a 500-nautical-mile circular orbit inclined 30° with respect to the equator will experience a nodal regression of 5.5° per mean solar day. Thus, every 65 days or so, its orbit plane will twist through one complete revolution of 360°.

For high-altitude satellites, Earth behaves more like a point mass, so their nodal crossing points precess at a slower rate. Specifically, at the GPS altitude 10,898 nautical miles above Earth, the nodal crossing point marches around the equator only 14° per year. This is a relatively small deviation compared with those experienced by low-altitude satellites, but, nevertheless, Earth's equatorial bulge creates the largest perturbation acting on the GPS satellites.

Figure 2.12 highlights another kind of perturbation induced by Earth's equatorial bulge. When we launch a satellite into an elliptical orbit, it does not travel along precisely the same ellipse over and over again. Instead, it moves along a spiraling trajectory resembling the one on the left-hand side of Figure 2.14. Notice now the initial semi-major axis line of the satellite's orbit is tilted to the right. Each time the satellite coasts up toward apogee, it traces out a slightly different elliptical orbit. Its line of apsides (semi-major–axis line) gradually rotates in the counterclockwise direction. The rate at which it rotates is denoted by $\dot{\omega}$.

Specifically, if a satellite is in a slightly elliptical 500-nautical-mile orbit with a 30° inclination, its line of apsides will rotate about 10° per mean solar day. If we launch a satellite into an

Figure 2.14 A satellite coasting around an elliptical orbit above a planet with an equatorial bulge will experience a systematic rotation of its line of apsides. The sketch on the left shows how that relentless rotation occurs. The graph on the right represents the apsidal rotation rates for a sampling of elliptical orbits all of which have 100-nautical-mile perigee altitudes. If a satellite's orbit is inclined 63.4° with respect to the equator, its apsidal rotation rate will always equal zero regardless of the eccentricity of its orbit or the size of the planet's equatorial bulge.

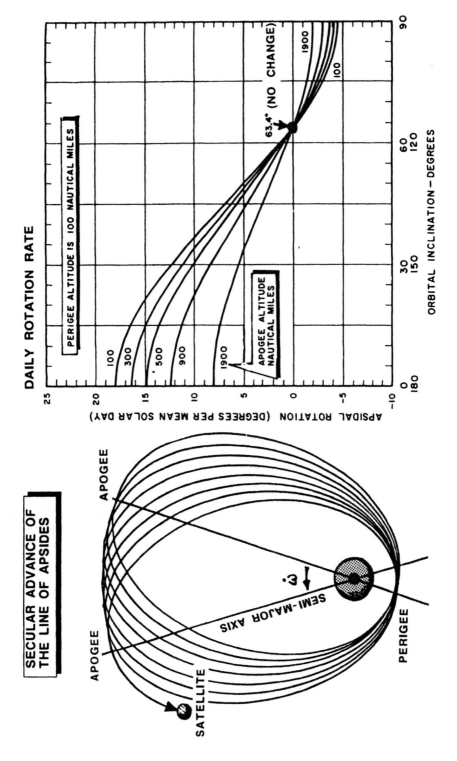

ASPIDAL ROTATION RATE

DAILY ROTATION RATE

SECULAR ADVANCE OF
THE LINE OF APSIDES

orbit with a 63.4° orbital inclination, its line of apsides will not rotate at all. This curious result can be verified by examining the graph in Figure 2.14 or by studying the equation for $\dot{\omega}$:

$$\dot{\omega} = \frac{(3/2)J_2 R_o \mu}{a^{3/2}(1 - e^2)^2} \left(2 - \frac{5}{2} \sin^2 i\right)$$

In this equation, constant J_2 defines the size of Earth's equatorial bulge. The other parameters have all been previously defined.

None of the multipliers outside the braces can ever have a zero value unless the earth is spherical. So $\dot{\omega}$ can equal zero if and only if

$$\left(2 - \frac{5}{2} \sin^2 i\right) = 0$$

Two different values for the inclination i satisfy this relationship.

$$i = 63.435°$$
$$i = 116.565°$$

The first of these two inclinations ($i = 63.435°$) is used effectively in connection with the Russian Molniya satellites. The second of the two inclinations ($i = 116.565°$) will be used by the Ellipso constellation of mobile communication satellites to provide spaceborne cellular telephone services for the entire globe.

3 THE ORBITAL ENVIRONMENT

In June 1665, as the Great Plague ran its deadly course, Cambridge University closed down and Isaac Newton set off for the peace and quiet of his rural English birthplace, where he would spend a year. This period was so rich in discovery that future historians would refer to it as the *annus mirabilis*, the "miraculous year."—*Jean-Pierre Maury (Newton: The Father of Modern Astronomy*,
Abrams, New York, 1992)

In October 1957, when the Russians hurled their first Sputnik into space, I was a junior in college enthusiastically chasing large swarms of elusive coeds all around a green, grassy campus. That next Friday afternoon, I hitchhiked 60 miles along a dusty country road to my hometown, Springfield, Kentucky, population 2000.

Early the next morning, I was walking down Main Street in front of Milburn's Shoe Shop when I spotted my boyhood friend, chubby, freckle-faced Johnny Hardin, who as always, had a big grin on his face. "Hey, Tommy," he called out, as he rushed across the street. "I just heard the Russians have a new artificial Earth satellite. They have one, do you think we should have one?"

I'm a little embarrassed to admit it now, but at the time, I couldn't think of a single reason why the United States should have its own satellite. But, before I could formulate any reply, Johnny Hardin answered his own question. "I don't think we should be willing to spend all that money," he concluded. "After all, there's *nothin'* out there!"

THE BENEFICIAL PROPERTIES OF SPACE

Neither of us knew it at the time, but Johnny Hardin was precisely right. There is nothin' out there. In fact, there are three different kinds of nothin' out there and those three kinds of nothin' are already enriching our lives in a hundred dozen different ways.

No Gravity

The first important kind of nothin' in outer space is *no gravity*, or the so-called microgravity. The microgravity of space arises because an orbiting satellite "falls" around its orbit with only the smallest counterforces to resist its forward motion. Surprisingly, microgravity was being used on Earth to make useful products several decades before the start of the Civil War.

In the early 1800s alert munitions makers noticed that when raindrops descended in weightless free fall, they automatically formed into nearly perfect spheres. That simple observation led to the construction of tall, slender "drop towers" into which molten globules of metal were released. As they descended toward the bottom of the tower in weightless free fall, those metal globules solidified into canon grapeshot, buckshot, and crude ball bearings. Similar drop towers are still being used today to manufacture metal spheres and in the production of the tiny glass beads found on reflecting highway markers.

Move from the surface of Earth up into outer space and the "gravitational" forces acting on your body will decrease by a factor of a million. The low-level, long-duration microgravity that can be routinely achieved aboard orbiting satellites will allow far-sighted entrepreneurs to manufacture nearly perfect crystals, higher-temperature turbine blades, high-strength permanent magnets, and powerful new pharmaceuticals—all of which may have revolutionary impacts on Earthbound consumer products and production techniques.

No Air

The second important kind of nothin' found in outer space is *no air*—or the so-called hard vacuum. Here on Earth, vacuum technology has been responsible for the creation of many profitable new industries. Every incandescent light bulb houses a man-made vacuum. So does the picture tube in every color television set. Thermos bottles, refrigeration equipment, the freeze drying of foods, the welding of dissimilar metals, and the manufacture of powerful new computer chips all rely on vacuum technology.

Three hundred miles above your head, the ambient air is only about one-trillionth as dense as the air now filling your lungs. The vacuum of space allows engineers and designers to construct large, floppy structures that can be inflated in orbit with surprisingly small amounts of gas.

In the 1960s, when a team of aerospace professionals inflated a giant Echo balloon in a New Jersey dirigible hanger, they used 80,000 pounds of gases to stretch it into a spherical shape. But, when that same balloon was later inflated in the natural vacuum of space, only 30 pounds of inflating gasses were required.

In the near future, various other types of inflatable structures may take advantage of the natural vacuum in space: orbiting astronomical observatories, semirigid habitation modules, and perhaps even donut-shaped space stations. Vacuum welding devices, and unenclosed klystrons for beaming energy to the ground or between orbiting satellites may also benefit from the extremely tenuous spaceborne atmosphere.

Nothin' to Block Your View

The third valuable kind of nothin' in outer space is *nothin' to block to your view*, or the so-called wide-angle view. Here on Earth, the wide-angle view has long been exploited in connection with radio and television transmitting towers, aerial crop surveys, and mapmaking operations, revolving restaurants, and military observation posts.

Move from the surface of Earth upward into outer space and the fraction of Earth you will be able to see at any given moment will increase by a factor of 500,000 (compared with having your eyes six feet off the ground). The wide-angle view from space has turned out to be by far its most valuable environmental property. Orbiting meteorological satellites, communication platforms, and large constellations of navigation satellites all take advantage of the wide-angle view from space. An unobstructed view also benefits space telescopes and Earth resources satellites—which can be used to monitor mineral deposits, the health of food crops, the growth of urban areas, and global sources of pollution.

At this moment, commercial communication satellites are beaming hundreds of thousands of telephone calls and several hundred color television shows to and from remote corners of the globe. Religious sermons, stock market quotes, cash register receipts, even whole newspapers are being relayed through orbiting satellites. *USA Today*, for instance, is assembled at one central location and then beamed to remote printing centers scattered across the nation. This approach provides simultaneous nationwide (and worldwide!) distribution, effortlessly—at affordable rates.

The Emerging Benefits of Space Technology to the Residents of Springfield, Kentucky

Johnny Hardin still lives in Springfield, Kentucky, but he no longer regards space as an unexploitable domain. Cable TV shows come into his home from orbiting satellites along with the weather maps he sees on the evening news. Moreover, when Johnny Hardin reads *USA Today* or the *Wall Street Journal*, he knows that he is sampling modern space-age technology.

The proprietors of Milburn's Shoe Repair Shop share their building with a new business. It is called B&E (Bates and Edleman) Television. The owners sell and service inexpensive satellite dishes to their many satisfied customers who enjoy multichannel TV while they farm the rich bottomlands surrounding Springfield, Kentucky, on every side.

THE SALIENT CHARACTERISTICS OF THE ORBITAL ENVIRONMENT

The primary purpose of this chapter is to explore some of the properties of the orbital environment that surrounds our home planet—especially those properties that affect orbiting satellites in detrimental or beneficial ways. Specific topics include Earth's atmosphere, its gravitational field, and its magnetic properties. The benefits and the hazards of the natural meteoroids and the artificial space debris fragments now whizzing around Earth are also briefly reviewed.

These special environmental properties strongly influence the art and science of orbital mechanics, particularly orbital altitude selection trades, spacecraft subsystem design, on-orbit lifetime predictions, shielding calculations, and the like.

EARTH'S ATMOSPHERE

Continuous collisions with the ambient air molecules in the upper atmosphere cause an orbiting satellite to lose energy and spiral down toward Earth. Eventually, if no compensating maneuvers are executed, it will plunge through the atmosphere when enough of its orbital energy has been eaten away.

The instantaneous drag force acting on a satellite can be expressed as follows:

$$D = \tfrac{1}{2} C_D A \rho V^2$$

In this equation, C_D represents the satellite's drag coefficient, A denotes the cross-sectional area it is presenting to the onrushing air, ρ is the density of the ambient atmosphere, and V represents the magnitude of the satellite's current velocity vector.

At first, the drag force is relatively weak and it reduces the orbital altitude of the satellite rather slowly, but when it encounters the denser layers of air at lower altitudes, its rate of decay begins to increase. A decay-rate history for a typical satellite is highlighted in Figure 3.1. In this case, the satellite starts out in a 300-nautical-mile circular orbit, but after about 700 days, it loses altitude so rapidly it slams into the ground within a few more days.

Two important factors influence the orbital decay rate of a satellite traveling around a low-altitude circular orbit:

1. Its ballistic parameter $(W/C_D A)$
2. Its initial orbital altitude

A satellite with a large ballistic parameter (bowling ball) will decay more slowly than one with a smaller ballistic parameter (volleyball). This is true because the bowling ball, with its higher density, has more inertia per unit area to resist the retarding force induced by atmospheric drag. Consequently, it will decay at a slower rate.

If a satellite is launched into a low-altitude orbit, it will spiral down toward Earth much faster than one in a higher-altitude orbit. This is primarily because Earth's atmosphere is denser at lower altitudes. The relentless increase in the atmospheric

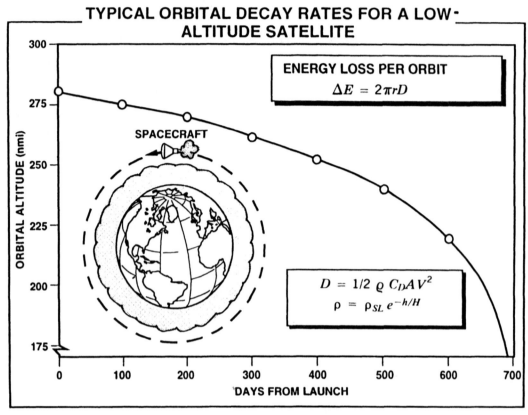

TYPICAL ORBITAL DECAY RATES FOR A LOW-ALTITUDE SATELLITE

ENERGY LOSS PER ORBIT

$$\Delta E = 2\pi r D$$

SPACECRAFT

$$D = 1/2\, \varrho\, C_D A V^2$$
$$\rho = \rho_{SL}\, e^{-h/H}$$

ORBITAL ALTITUDE (nmi)

DAYS FROM LAUNCH

Figure 3.1   As this curve indicates, a satellite traveling around a circular orbit decays slowly at first; then, as it moves down into the denser layers of Earth's atmosphere, it decays at a more rapid rate. In this case, the satellite is in a 300-nautical-mile circular orbit and 500 days go by before its altitude is reduced to 200 nautical miles. At that lower altitude, the atmosphere is considerably denser, so Earth impact occurs less than 200 days later.

density at lower altitudes occurs because the air above compresses the air below. To a first approximation, the density of the atmosphere is a negative exponential function of altitude. Thus, we can express the local atmospheric density ρ at any altitude *h* as:

$$\rho = \rho_{SL} e^{-h/H}$$

In this equation, ρ_{SL} represents the density of air at sea level (roughly, 14.7 pounds per square inch), and H denotes the atmosphere's "scale height" of 4.8 miles.[1]

At sea level, the density of the atmosphere is approximately 1/840 times the density of tap water. Thus, we see that our atmosphere is surprisingly thin and tenuous. Moreover, it decreases at an extremely rapid rate with increasing altitude. Approximately 99 percent of it lies below an altitude of 20 nautical miles.

The drag on an orbiting satellite causes continuous orbit decay and spacecraft heating—especially for high-energy space vehicles returning from the Moon or other deep-space locations. Of course, atmospheric drag also can be helpful for space exploration. It allows us to execute aerobraking maneuvers such as the ones used in connection with Project Apollo. It also fosters the parachute recovery of deep-space probes, film packets ejected from reconnaissance satellites, and emergency rescue operations.

EARTH'S GRAVITATIONAL FIELD

To a first approximation, our beautiful blue planet—third rock from the Sun—is a massive sphere 6888 nautical miles in diameter. Three-fourths of its surface is dimpled with relatively shallow puddles of fresh or salty water and it is wrapped in a thin film of breathable air. Our Earth weighs about 13 million-billion-billion pounds. By virtue of its large mass, our home planet has a surface gravity strong enough to pull any loose objects near its surface vertically downward at an acceleration rate of 32.174 ft/sec^2. Hurling a one-pound object from the surface of Earth onto an escape parabola requires an energy expenditure of 20 million foot-pounds.

As Isaac Newton proved mathematically, objects coasting along gravity-induced trajectories in the vicinity of Earth trace out rather precise conic sections: circles, ellipses, parabolas,

---

[1]If Earth's atmosphere could be perfectly characterized by the negative exponential function, its density at the top of any 4.8-mile vertical column of air always decreases by 64 percent compared with its density at the bottom.

and hyperbolas. These simple two-body orbits are subjected to noticeable perturbations induced by atmospheric drag, solar-radiation pressure, the gravitational fields of the Sun and the Moon, and the subtle gravitational influences of the higher-order shape terms of Earth—the largest of which is Earth's second potential harmonic caused by its rather pronounced equatorial bulge.

Earth's atmosphere is held in place by the force of gravity, which also acts as a powerful "lens" to attract meteoroids (cometary fragments and small asteroids) some of which would, otherwise, miss Earth.

In 1687, Isaac Newton proved that Earth has an equatorial bulge several miles high. His clever line of reasoning focused on the individual water molecules suspended on the surface of the ocean. These molecules are pulled down toward the center of Earth by gravity, while they are being hurled rapidly outward by the centrifugal force created by Earth's rotation around its north–south axis of spin.

Newton reasoned that the surface of the ocean should automatically assume an "equipotential" surface along which the vector sum of these two forces is everywhere in balance. Then he used integral calculus to prove that the equipotential surface it would assume is an oblate spheroid.

Water obviously has no rigidity to resist the influence of these two forces, but Isaac Newton had no way of knowing whether, over eons of time, the solid portions of Earth were rigid enough to put up long-term resistance. Later, however, he conjectured that both Earth and water must be similarly affected; otherwise, the wall of water—which he determined was several miles high—would permanently inundate all the land masses lying on or near the equator.

The equatorial bulge constitutes, by far, Earth's largest and most influential departure from a simple spherical shape. However, geophysicists have, for years, carefully measured the perturbations acting on orbiting satellites to determine the values for at least 180 higher-order-shape terms defining, in exquisite detail, the precise gravitational field of our planet. For many orbital mechanics calculations, these higher-order-shape terms can be safely ignored. But for precise trajectory simulations—such as those required for the targeting of intercontinental ballistic missiles flying silo-busting missions—all 180 of them are sometimes required.

In the 1960s, the engineers who masterminded the Transit navigation satellites needed to devise an inexpensive way to orient each satellite so its navigation antennas would point always toward the center of Earth. A continuous Earth-seeking orientation was required for mission success because each of the Transit satellites was required to blanket the entire visible portion of Earth with radio frequency transmissions.

Eventually, a way to achieve the necessary attitude control was perfected that used the natural gravity-gradient forces induced by Earth's inverse square gravitational field. First, the mission planners launched each Transit satellite into a 540-nautical-mile polar orbit. Then they commanded it to point its navigation antennas toward the center of Earth. Finally, they extended a 50-foot telescoping boom vertically upward away from the center of Earth.

The gravitational force pulling on the lower portion of such an elongated satellite is stronger than the gravitational force pulling on its upper portion. This subtle "gravity gradient" tends to keep the satellite in an Earth-seeking orientation as it travels around Earth.

Any small disturbing torque acting on a gravity-gradient stabilized satellite makes it oscillate about the vertical like a big, long, rigid pendulum. Fortunately, these small oscillations can be damped out by electrical hysteresis loops powered by electrical energy from the Sun.

On some missions, the space shuttle engineers take advantage of gravity-gradient stabilization to achieve an Earth-seeking orientation for the shuttle orbiter as it coasts around its orbit. On such a mission, the orbiter is purposely flown so its nose or its tail is continuously pointing toward the center of Earth. No telescoping booms are required in this case; the natural elongated shape of the orbiter induces all necessary gravity-gradient stabilization forces.

THE MAGNETIC FIELD SURROUNDING EARTH

Throughout the last 800 years, Earth's natural magnetic field has helped compass-equipped mariners find their way across vast uncharted seas. By the twelve century A.D., they somehow learned that lodestones (natural occurring magnets), when stuck through a small bit of floating cork, would seek out the northward direction.

Earth's natural magnetic dipole is roughly aligned with its north–south axis of spin. At sea level, its magnetic field strength is about 0.5 gauss, but with increasing altitude, that field strength decreases at a rate that is approximately proportional to the inverse cube of the radial distance from the center of Earth.

Earth's magnetic dipole points in the general north–south direction, but it is imprecisely aligned with Earth's spin axis, and its center does not coincide with the center of mass of Earth. A 212-mile mismatch between the geographic center and the magnetic center of Earth induces a permanent dip in the lower Van Allen radiation belt known as the "South Atlantic Anomaly." Low-altitude satellites traveling through the South Atlantic Anomaly (near the East coast of South America) encounter far more hazardous radiation than they otherwise would.

Earth's magnetic field can create steady torques on charged satellites or those that carry activated electromagnets. Those torques can be an undesirable nuisance to mission planners. But for some missions, carefully controlled torques can be purposely induced to provide magnetic momentum dumping for the satellite's reaction wheels (heavy spinning flywheels). This widely used technique helps a satellite maintain relatively precise attitude control without expending valuable propellants during extended portions of its on-orbit life.

Contour lines defining the field strength of the magnetic dipole at the surface of Earth are presented in Figure 3.2. The South Atlantic Anomaly is situated near the center of this graph. Above that region the energetic swarms of protons and electrons in the lower Van Allen Radiation Belt swoops down closer to the surface of Earth. Mission designers must carefully take this extra radiation into account to minimize any damage to their low-altitude satellites. Damage reduction can be achieved by structuring the mission so the spacecraft does not fly near the South Atlantic Anomaly, by careful component shielding, or by designing the satellite's subsystems so they can withstand the extra radiation they will encounter.

Gravity-gradient stabilization is usually impractical for medium-altitude and high-altitude satellites because the gravity-gradient torques fall off so rapidly with increasing altitude. Fortunately, Earth's magnetic field can be harnessed to help maintain an Earth-seeking orientation without using propellants.

Newton's third law (action-reaction) is routinely employed to control the attitude of an orbiting satellite. This is accomplished by using the satellite's electrical power supplies to adjust the rotation rates of a set of heavy spinning disks called

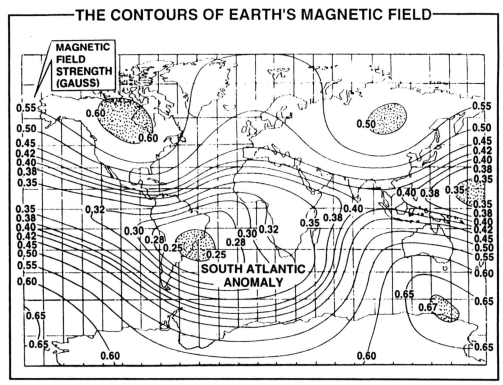

Figure 3.2 The natural magnetic field on the surface of Earth averages about 0.5 gauss. However, in a large region called the South Atlantic Anomaly just below the nose of Brazil, the lines of constant field strength are weaker and the Van Allen Radiation Belts dip down closer to Earth. Low-altitude satellites flying through the South Atlantic Anomaly experience much greater damage from Van Allen radiation than those flying at the same altitude over other portions of Earth.

"reaction wheels" (or "momentum wheels"). Whenever an unwanted torque begins to disturb the attitude of the satellite, the spin rate of these reaction wheels is modified automatically to force it back into the desired orientation.

Generally speaking, over time, the satellite's reaction wheels tend to spin faster and faster, so they must be slowed down periodically. One way to slow them down is by firing a pair of small onboard rockets arranged in a couple to create a counter-torque. Unfortunately, this simple approach consumes precious onboard propellants, thus limiting the on-orbit life of the satellite.

However, an alternate approach has been perfected for slowing down the reaction wheels without using propellants.

This is accomplished by selectively activating three mutually orthogonal electromagnets, thus creating controlled counter-torques as the magnets cut across Earth's magnetic lines of flux. This approach does not use valuable propellants, but, of course, it does consume some of the electricity being generated by the satellite's solar arrays.

METEOROIDS IN SPACE

Naturally occurring meteoroids are stony, metallic, or soil-like particles careening through space at Earth escape velocities and beyond. Many meteoroids are associated with old comet trails or the belt of asteroids orbiting the Sun mostly between Jupiter and Mars.

A stray meteoroid can puncture the walls of an orbiting satellite or an astronaut's space suit with potentially catastrophic results. Large, destructive meteoroids slam into the ground infrequently, but when such an impact does occur, it can be far more destructive than the largest man-made thermonuclear explosions.

The Meteor Crater in Arizona, for instance, which is 4150 feet across and 570 feet deep, is believed to have been created by the hypervelocity impact of a metallic meteoroid approximately 100 feet in diameter. Recent scientific studies indicate that the average American is more likely to die due to an encounter with a stray meteoroid than in the crash aboard a commercial jetliner. During most centuries, no American deaths at all will occur due to meteoroids hitting Earth. But a large, infrequent meteoroid impact in a heavily populated area could kill thousands or even hundreds of thousands in a second or two. Averaged over several centuries, a few hundred Americans per year will probably be killed by meteoroid impacts. This could easily exceed the 100 to 200 Americans killed in commercial jetliner crashes in an average year.

Meteoroids orbit the Sun along gravitationally induced trajectories. Most of them are randomly distributed in our sector of the solar system, but thick swarms of meteoroids are associated with old comet trails curling around the Sun. Each year, as Earth sweeps through these semipermanent trails of space debris, alert individuals looking up at the night sky can see a "shooting star" every minute or two. Those comet-trail encounters occur at predictable intervals. Sometimes newspapers print announcements alerting the public when a particularly impressive meteor shower is going to occur.

Meteoroids orbit the Sun along relatively precise elliptical free-fall trajectories. But, when one of them happens to pass close to a planet, the resulting swing-by can suddenly jerk it onto an escape hyperbola or onto a completely different elliptical orbit around the Sun. Consequently, the long-term trajectories of meteoroids, large or small, are essentially unpredictable.

A crude statistical histogram defining the velocity distribution of sporadic meteorites is presented in Figure 3.3. The average speed at which they enter Earth's atmosphere is about 12 miles per second. This is roughly 1.7 times faster then the

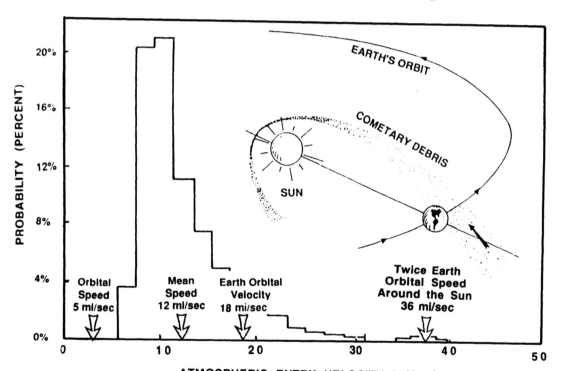

PROBABILITY VELOCITY DISTRIBUTION FOR SPORADIC METEOROIDS

Figure 3.3 This probability histogram defines the velocity distribution for the sporadic meteoroids entering Earth's atmosphere. Their average entry velocity is about 12 miles per second. A small number of them encounter Earth at a velocity of about 18 miles per second, which equals the speed at which Earth travels around the Sun. A still smaller number come into the atmosphere at 36 miles per second because they orbit the Sun in the retrograde orbits and meet Earth head on.

reentry speed of the Apollo capsules that brought the astronauts back from the Moon. A few of the sporadic meteoroids travel around the Sun along retrograde orbits. Their reentry speeds are about 36 miles per second—or twice the speed at which Earth is orbiting the Sun.

Meteoroids usually experience a fiery death when they breach Earth's atmosphere. But a few large ones and a multitude of small ones survive that drag-induced baptism of fire. The outer layers of large-hypervelocity meteoroids are singed, melted, or vaporized during reentry, but their inner cores sometime survive to reach the ground.

The tiny fragments that survive their journey through Earth's atmosphere are called micrometeorites. They are slowed down to relatively low velocities in the tenuous, high-altitude layers of the atmosphere. Slight surface-layer melting usually causes them to assume a pitted, roughly spherical shape.

Naturally occurring meteoroids can damage orbiting satellites and erode their exposed surfaces. However, in the low-altitude flight regime, man-made objects careening through space pose a much greater hazard to orbiting satellites.

Man-Made Space Debris

Satellites, old and new, spent rocket casings, paint flakes, and the jagged fragments created by inadvertent or preplanned explosions in space make up most of the man-made space debris now orbiting Earth. Most of these debris fragments occupy nearly circular orbits barely skimming over the atmosphere, but smaller concentrations also exist at the 12-hour semisynchronous and the geosynchronous altitudes.

Large orbiting objects can collide with bomblike energy, but debris fragment as small as a pinto bean can pose a serious hazard to an orbiting satellite—especially a large inflatable structure or a manned space station. Gradually, as these hazards are becoming more apparent to spacefaring nations, measures are being taken to limit the steady accumulation of space debris.

One relatively easy way to reduce the population of debris fragments is to eliminate spaceborne explosions. To date, approximately one hundred explosions are known to have occurred in space, most of which have created at least several dozen new debris fragments. Many of these explosions occurred when unused propellants trapped in old upper-stage rockets were heated by the sun or mixed inadvertently.

Other spaceborne explosions have been triggered on purpose by U.S. and Soviet military specialists who were conducting tests to make sure their "killer satellites" would operate properly in space. In preparation for such a test, Soviet rocketeers would launch a target satellite into orbit around Earth. Then they would chase it down with a killer satellite. When the killer satellite got close enough, it would blow itself up to engulf the target in a deadly cloud of space debris.

U.S. military space commanders operate 30 facilities equipped with optical and radar sensors capable of tracking surprisingly small fragments hurling through space. Their special instruments are currently being used to track about 7000 fragments four inches in diameter or larger. Another 17,000 objects ranging between 0.4 and 4 inches in diameter are currently coasting around Earth in low-altitude orbits.

Because of atmospheric drag, space debris fragments are constantly plunging back toward Earth, but, of course, new ones are also being lofted into space. Current growth rates average about 175 trackable fragments per year. Man-made objects totaling 6 million pounds are estimated to be orbiting Earth. Industry experts predict that this cloud of fragments will grow to 24 million pounds in the next 15 years.

At any altitude below 500 nautical miles, space debris fragments sent up from Earth pose a greater hazard to orbiting satellites than naturally occurring asteroids and cometary fragments. This estimate is based on the assumption that a collision with any object larger than 0.4 inch in diameter poses a hazard to an orbiting satellite.

Because of their savage velocities, solid objects traveling around Earth can cause catastrophic damage to any unprotected satellite they happen to encounter. The maximum kinetic energy exchange during such a collision is graphed in Figure 3.4 for low-altitude satellites and those hovering above Earth's equator at the geosynchronous altitude. The angle of intersection between the two orbiting fragments is plotted along the horizontal scale. The vertical scale represents the maximum kinetic energy exchange resulting from a solid collision.

As the curve at the top of Figure 3.4 indicates, any pair of one-pound fragments in low-altitude orbits whose trajectories intersect at a 15° angle or greater could, on impact, generate a collision energy equaling the detonation of two pounds of TNT.

COLLISION ENERGY FOR ORBITING OBJECTS

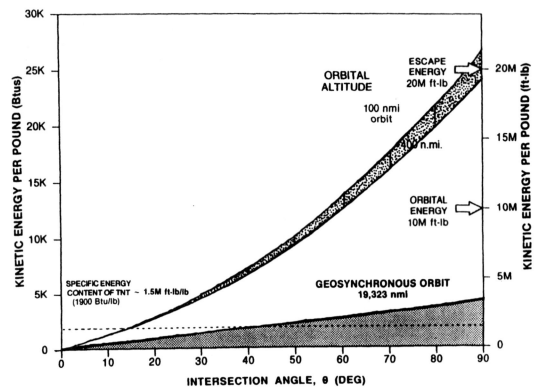

Figure 3.4 A solid collision between two one-pound satellites slamming into one another at a 15° intersection angle in low-altitude circular orbits could equal the energy resulting from the detonation of a similar amount of TNT. A similar collision between geosynchronous satellites typically would be less energetic because geosynchronous satellites travel slower and because they all occupy essentially the same plane.

Collisions between satellites at the geosynchronous altitude tend to be somewhat less energetic. Their lower destructive potential results because geosynchronous satellites travel slower and they are all orbiting in essentially the same plane. Nevertheless, a collision at the geosynchronous altitude could still involve a destructive energy exchange which, in turn, could create a much larger number of smaller, but still hazardous debris fragments.

In the later 1980s, a team of researchers at General Motors conducted a series of tests to measure the destructive effects of high-velocity projectiles slamming into a solid surface. In one widely reported experiment, they fired a quarter-ounce plastic cylinder into a 25-pound slab of aluminum at a velocity of 5.3 miles per second.[2] When the test was over, the engineers estimated that aluminum fragments totaling 1.1 pounds had been ejected from the aluminum block. Thus, the fragments of aluminum that were ejected outweighed the plastic projectile by a ratio of 70 to 1.

In 1986, a European Ariane upper stage coasting around a retrograde Sun-synchronous orbit suddenly exploded, thus creating nearly 3000 new space debris fragments. Initially, those fragments formed a thin belt passing in the vicinity of both the North Pole and South Pole. Within four years, however, that thin belt had spread out into a broad swarm of fragments enveloping Earth. Over time, the nodal crossing points of the individual fragments were perturbed differentially by Earth's equatorial bulge because the force of the explosion drove them into slightly different orbits.

Space debris fragments are constantly plunging back into the atmosphere, but they do not pose much of a hazard to the people on Earth. Many of them burn up before they reach the ground; others impact harmlessly in the oceans or in sparsely inhabited regions. Small portions of Earth are thickly populated with human beings, but, on average, we are only 27 tiny little bodies per square mile. Consequently, we are not often hit by hazardous objects falling from the sky—natural or man-made.

However, one relatively rare type of satellite can pose a substantial hazard to the inhabitants of our planet. Russian scientists have orbited about 40 radioactive payloads, most of which are still in orbit. At the end of their useful on-orbit lives, most of these potentially hazardous satellites are successfully boosted into higher altitude orbits that will keep them in space until their radioactivity has been largely depleted. However, even if a radioactive satellite reaches a higher orbit, a chance collision with a large debris fragment could cause part of it to reenter the atmosphere, thus scattering a broad swath of hazardous material over the ground.

A different scenario caused Cosmos 954 to slam into North America in the late 1980s. When Cosmos 954 reached the end of its useful life, Russian rocket scientists commanded it to boost itself into a higher orbit. Unfortunately, its reboost rockets failed to operate, and it continued to spiral down toward Earth. When it plunged into the atmosphere, it fell apart and

---

[2]An impact between two low-altitude satellites intersecting at a 60° intersection angle would duplicate the impact velocity in the General Motors test.

scattered radioactive debris over a rugged region in Northern Canada roughly the size of Delaware. Cleaning up the resulting radioactive contamination cost the Canadians 10 million U.S. dollars. The Soviets, who admitted that they caused the problem, paid half. The Canadians shelled out the other $5 million.

4 POWERED FLIGHT MANEUVERS

Philosophy is such an impertinently litigious lady that a man
had as good be engaged in lawsuits as to have do with her.
—*Isaac Newton in a letter to his friend Edmund Halley, June 20, 1687*

In space, velocity is the name of the game. If we can execute a
sequence of powered flight maneuvers to adjust the velocity of
a spaceship with sufficient precision, we can cause it to coast to
almost any desired location in space.

In this chapter, we will discuss the proper ways to carry out
various powered flight maneuvers including the classical
Hohmann transfer maneuver and the pure plane-change maneuver. We will also review the bielliptic transfer, on-orbit phasing and rendezvous, deorbit maneuvers, and the precise
planetary swing-bys that are rapidly opening up the solar system for efficient exploration by unmanned deep sea probes.

THE CLASSICAL HOHMANN TRANSFER MANEUVER

In 1925, the German engineer Walter Hohmann proposed a
specific sequence of maneuvers he believed would turn out to
be the optimal way to transfer a space vehicle from one co-
planar circular orbit to another. In Hohmann's approach, the
space vehicle would employ two separate impulsive rocket
burns parallel to the existing velocity vectors of the two circular orbits separated by a 180° coasting arc consisting of a
bitangent ellipse. The diagram in the upper left-hand corner of
Figure 4.1 depicts the geometrical properties of this two-burn
Hohmann transfer maneuver.

Notice that the transfer ellipse does not pierce either of the two co-planar circular orbits. It is tangent to the low-altitude orbit at perigee and it is tangent to the high-altitude orbit at apogee. Walter Hohmann was convinced that this two-burn impulsive sequence was optimum because no energy is wasted during either of the burns—both of which are executed parallel to existing velocity vectors.

For many years, nearly everyone in the aerospace industry believed that Hohmann's impulsive maneuver sequence was the most fuel-efficient approach. However, they later learned that a more complicated sequence of maneuvers involving *three* rocket burns, in some cases, could reduce the total ΔV budget. That approach, which is called the "bielliptic transfer," is discussed later in this chapter.

The magnitudes of the two separate impulsive velocity increments needed for various co-planar Hohmann transfer maneuvers are graphed in Figure 4.1. The thick bandlike curves enclose the ΔV's associated with initial orbits ranging from 100 to 300 nautical miles altitude. The logarithmic scale running across the bottom of the figure denotes the altitude of the final destination orbit (which is also assumed to be circular). The scale on the left-hand side represents the various velocity increments used in executing the indicated maneuvers.

Notice that one of the two vertical lines in the middle of Figure 4.1 corresponds to a co-planar geosynchronous destination altitude 19,300 nautical miles high. In this case, the first rocket burn requires a velocity increment (ΔV_1) of 7800 feet per second and the second burn requires a velocity increment (ΔV_2) of 4800 feet per second. The sum of these two velocity increments ($\Delta V_T = 12,600$ feet per second) lies along the curve at the top of the figure.

Figure 4.1 In 1925, the German engineer Walter Hohmann conjectured that the best way to maneuver a satellite from a low-altitude circular orbit to a higher altitude circular orbit sharing the same plane was to make two posigrade burns separated by a 180° coasting arc. The three parametric curves running across this chart represent the ΔV associated with the first Hohmann transfer burn, the ΔV associated with the second burn, and the total ΔV. Notice that the total ΔV required to go from a tree-skimming circular orbit to a high-altitude circular orbit reaches its maximum when the final destination orbit is 43,000 nautical miles high. Both higher and lower destination orbits require smaller total ΔV's.

THE VELOCITY INCREMENTS NEEDED FOR TYPICAL IMPULSIVE CO-PLANAR HOHMANN TRANFER MANEUVERS

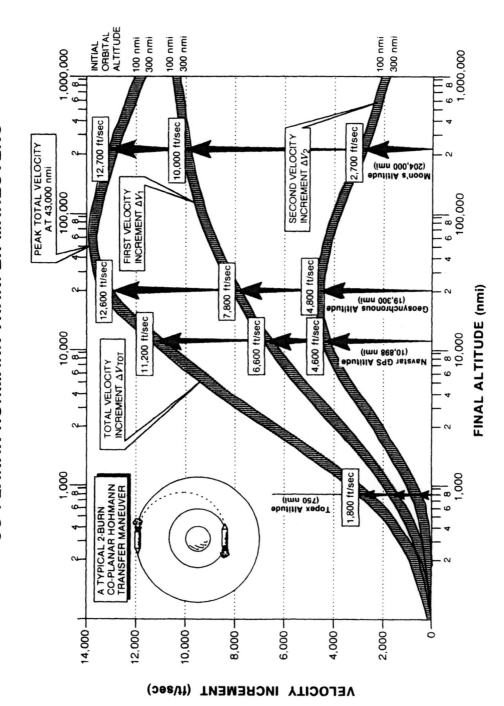

The most difficult co-planar circular orbit to reach from a low-altitude parking orbit turns out to be 43,000 nautical miles high. Its total velocity increment of 13,700 feet per second lies near the top of Figure 4.1. Strange as it may seem, a spaceship coasting around Earth in a 100-nautical-mile circular orbit can hurl itself onto an infinitely elongated ellipse (parabolic escape trajectory) with a total velocity increment ($\Delta V_T = 10,590$ feet per second)—about 3000 feet per second less than is needed to fly that same vehicle into a co-planar circular geosynchronous orbit.

At first thought, this may seem impossible because the total energy of any orbit—including this final destination orbit—is a function only of its semi-major axis—the larger its semi-major axis, the higher the energy of the orbit. Fortunately, this paradox can be explained by noting that *two rocket burns* are required to transfer the geosynchronous satellite into its final destination orbit. The propellants needed for the second burn must be carried upward against Earth's gravity. A careful energy-balance calculation that includes both the potential energy and the kinetic energy carried away by the perigee kick motor and the apogee kick motor's exhaust molecules shows that, as we would anticipate, energy is conserved.

LOW-THRUST AND MULTI-IMPULSE MANEUVER SEQUENCES

In the real world, of course, impulsive maneuvers do not exist. However, maneuver sequences involving rocket burns of reasonably brief duration (a few minutes total burn time) closely approximate the ΔV requirements depicted in Figure 4.1. Rocket burns of longer duration (hours, days, or weeks) involve losses ranging from a few percent up to 150 percent! These so-called "gravity losses" arise because the velocity is not all added in the proper direction and the rocket is constantly gaining altitude while a long-duration low-thrust maneuver is being executed. Consequently, a larger equivalent ΔV budget is required to hurl the spacecraft into the desired destination orbit. If it is equipped with an extremely low-thrust rocket, a space vehicle must sometimes spiral out to its destination over a period of several days or weeks.

Fortunately, a fuel-saving approach called the multi-impulse maneuver can be used instead of a continuous spiral—provided the low-thrust rocket can be extinguished and reignited repeatedly. When a low-thrust multi-impulse maneuver is being executed, the rocket is turned on and off over and over again near the perigees of a series of increasingly elongated transfer ellipses.

In such a maneuver sequence, the space vehicle executes a propulsive burn along a short thrusting arc, and then it shuts down its engines and coasts all the way around the new elliptical orbit. Then, when it arrives back at the perigee, it thrusts again to drive itself onto a slightly more elongated ellipse. This process is repeated as many times as necessary until the apogee of the greatly elongated transfer ellipse is tangent to the vehicle's final destination orbit.

When it reaches that apogee altitude, the vehicle can either execute one final propulsive burn to achieve circularization— or it can use another series of shorter multi-impulse burns separated by 360° coasting arcs. This rather complicated sequence of maneuvers takes longer, but it can provide substantial reductions in propellant consumption compared with a single long burn along a spiraling trajectory executed with the same low level of thrust.

PURE PLANE-CHANGE MANEUVERS

Some missions require pure plane-change maneuvers in which a space vehicle is switched from one orbit plane to another with no simultaneous change in altitude. In the simplest and most common case, the vehicle remains at the same altitude with a change in its orbital inclination. For small inclination changes, this is accomplished with a single impulsive rocket burn executed over the equator parallel to the local horizon and essentially perpendicular to the vehicle's current velocity vector.

Figure 4.2 highlights the geometry of a pure plane-change maneuver. The space vehicle is already in the "old orbit" and it must end up in the "new orbit," both of which are circular orbits sharing the same altitude. As the vector diagram at the bottom of Figure 4.2 indicates, the magnitude of the velocity increment required to close the vector diagram can be obtained from the following trigonometric relationship:

$$\sin \frac{\Delta\theta}{2} = \frac{\Delta V/2}{V_{\text{CIRC}}}$$

ΔV's REQUIRED TO MAKE VARIOUS PLANE CHANGES

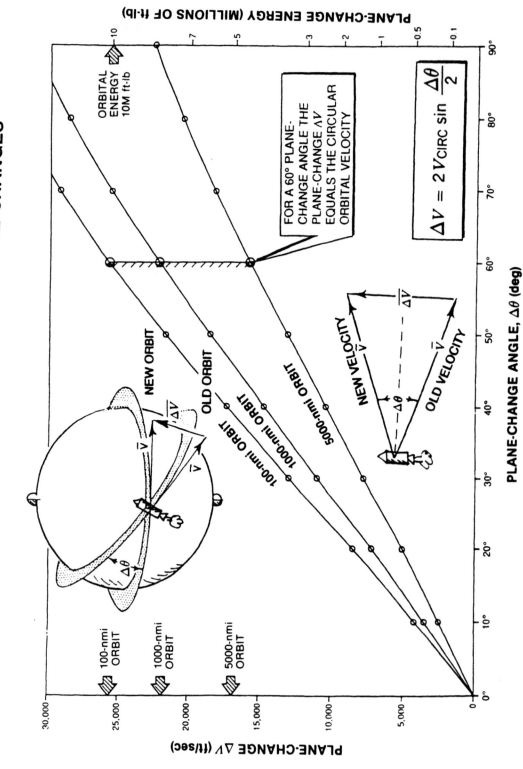

$$\Delta V = 2V_{CIRC} \sin \frac{\Delta \theta}{2}$$

FOR A 60° PLANE-CHANGE ANGLE THE PLANE-CHANGE ΔV EQUALS THE CIRCULAR ORBITAL VELOCITY

ORBITAL ENERGY 10M ft-lb

NEW ORBIT

OLD ORBIT

5000-nmi ORBIT

1000-nmi ORBIT

100-nmi ORBIT

NEW VELOCITY

OLD VELOCITY

100-nmi ORBIT

1000-nmi ORBIT

5000-nmi ORBIT

PLANE-CHANGE ENERGY (MILLIONS OF ft-lb)

PLANE-CHANGE ANGLE, Δθ (deg)

PLANE-CHANGE ΔV (ft/sec)

where θ is the required inclination change, ΔV is the magnitude of the velocity increment to be added, and V_{CIRC} is the circular orbital velocity of both the old and the new orbits.

Solving for the required increment (ΔV) gives the desired equation:

$$\Delta V = 2 V_{CIRC} \sin \frac{\Delta\theta}{2}$$

The three parametric curves in Figure 4.2 were constructed using this simple relationship. The curve at the top of the figure defines the ΔV's required for plane-change angles of various magnitudes starting and ending in a 100-nautical-mile circular orbit. The other two curves define the velocity increments required to make pure plane changes of various sizes in 1000- and 5000-nautical-mile circular orbits.

Notice that a plane change of a given magnitude always requires less ΔV if the satellite is in a higher-altitude orbit. This is true because, at higher altitudes, the satellite travels at a lower velocity around its circular orbit. Consequently, a plane change of a fixed angular magnitude is easier to execute.

Notice that at any altitude, a 60° plane change always requires an impulsive velocity increment (ΔV) that equals the circular orbital velocity at that same altitude. If you encounter a situation in which you must execute a 60° plane change in a low-altitude orbit, you may as well start over again and launch a new satellite.

Figure 4.2 Pure plane changes in space are energy-wasting maneuvers. In executing such a maneuver, the rocket is oriented essentially perpendicular to the direction of travel and parallel to the local horizon. As these parametric curves indicate, the ΔV's associated with small plane changes are essentially linear functions of the total plane-change angle. For higher-altitude circular orbits, the ΔV required to execute a plane-change maneuver of a given magnitude is always smaller because high-altitude satellites travel around their orbits at slower speeds.

COMBINED PLANE CHANGES

If a space cadet decides to fly from one circular orbit to another while executing a plane-change maneuver, several different approaches are possible. One choice would be to execute a pure plane-change maneuver followed by a two-burn Hohmann transfer to handle the necessary change in altitude. This approach is appealing simple, but it does not result in the minimum total ΔV.

Another approach would involve a two-burn Hohmann transfer sequence followed by a pure plane-change maneuver executed at the destination altitude. This approach lowers the total ΔV budget, but it is not the optimum choice. A third possibility would be to employ a two-burn Hohmann maneuver in which the necessary plane change is combined with the second Hohmann transfer burn. This approach is not optimum, but in some cases, it can provide substantial ΔV savings compared with the other two approaches that have been discussed so far.

In the real world, the optimum approach usually consists of splitting the plane change optimally between the two burns in a classical Hohmann transfer maneuver sequence. In this case, our clever space cadet would make a small portion of the required plane change in conjunction with the first (perigee) burn, and then remove the remainder in conjunction with the second burn at the apogee of the transfer ellipse.

A maneuver sequence involving an optimal plane-change split is depicted in Figure 4.3. In this case, a commercial communication satellite is launched due east out of Cape Canaveral into a 100-nautical-mile parking orbit with a 28.5° orbital inclination.[1] Once the satellite enters its 100-nautical-mile parking orbit, it coasts until it is directly over the equator. Then it executes a two-burn Hohmann transfer sequence with an optimal plane-change split. This sequence of maneuvers car-

---

[1] The 28.5° orbital inclination arises naturally because it equals the latitude of Cape Canaveral.

Figure 4.3 This graph highlights the optimal plane-change split associated with a combined orbit-raising/plane-change maneuver in which a satellite must be moved from a 28.5° circular orbit at an altitude of 100 nautical miles to an equatorial geosynchronous orbit. In this case, the minimum total ΔV is associated with a plane-change split of 2.16° at the perigee burn and 26.34° at the apogee burn.

ORBITAL GEOMETRY FOR THE HOHMANN TRANSFER MANEUVER TO GEOSYNC

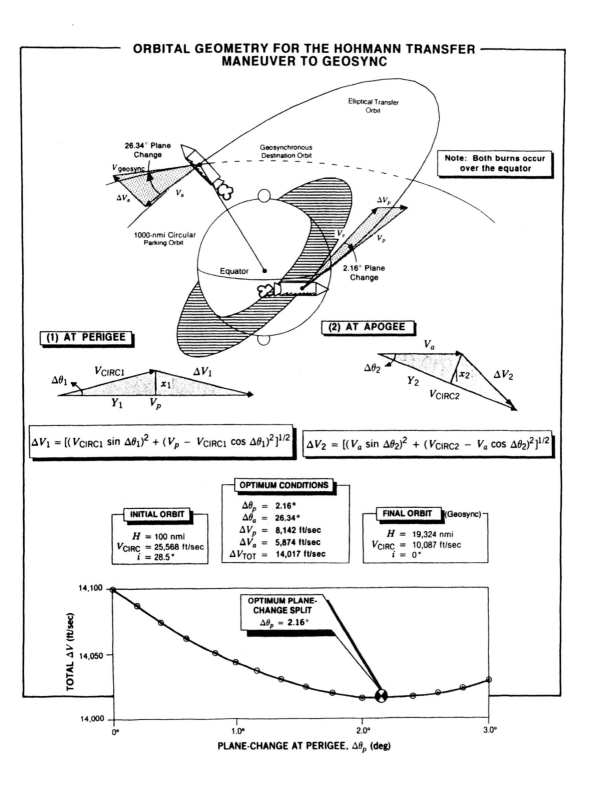

(1) AT PERIGEE

$$\Delta V_1 = [(V_{CIRC1} \sin \Delta\theta_1)^2 + (V_p - V_{CIRC1} \cos \Delta\theta_1)^2]^{1/2}$$

(2) AT APOGEE

$$\Delta V_2 = [(V_a \sin \Delta\theta_2)^2 + (V_{CIRC2} - V_a \cos \Delta\theta_2)^2]^{1/2}$$

INITIAL ORBIT

H = 100 nmi
V_{CIRC} = 25,568 ft/sec
i = 28.5°

OPTIMUM CONDITIONS

$\Delta\theta_p$ = 2.16°
$\Delta\theta_a$ = 26.34°
ΔV_p = 8,142 ft/sec
ΔV_a = 5,874 ft/sec
ΔV_{TOT} = 14,017 ft/sec

FINAL ORBIT (Geosync)

H = 19,324 nmi
V_{CIRC} = 10,087 ft/sec
i = 0°

OPTIMUM PLANE-CHANGE SPLIT

$\Delta\theta_p$ = 2.16°

ries the satellite up to its geosynchronous destination orbit 19,300 nautical miles above the equator with the desired 0° orbital inclination. Of course, the sum of the two plane changes made in conjunction with the two Hohmann transfer burns must equal 28.5°.

The equations for the two ΔV's can be derived by using the Pythagorean theorem in conjunction with the two simple vector diagrams in Figure 4.3. The first ΔV equation applies to the perigee burn; the other one applies to the apogee burn. These vector diagrams can be used to determine the directions in which the two velocity increments must be added to achieve the desired results.

When the communication satellite is coasting around its 100-nautical-mile circular parking orbit, its velocity is 25,568 feet per second and its orbital inclination is 28.5°. After the first burn, which must be executed over the equator, the satellite coasts half way around Earth to the apogee of the transfer ellipse, which also lies over the equator. When it arrives at the apogee of its transfer ellipse, it has a velocity of 5874 feet per second and an inclination of 26.34°. At this point, it must make a second burn to increase its velocity to 10,087 feet per second while reducing its orbital inclination to 0°.

The curve in Figure 4.3 defines the total ΔV budgets that result from splitting the total 28.5° plane-change angle in various ways between the two maneuvers. If the entire plane change is made at apogee, the total ΔV budget turns out to be 14,100 feet per second.

Splitting the total plane-change angle optimally reduces this total ΔV budget by about 80 feet per second. The optimal plane-change split requires that 2.16° be removed in conjunction with the larger perigee burn. The remaining 26.34° is removed in conjunction with the smaller apogee burn.

Splitting the plane change optimally can produce substantial performance gains. Sometimes, however, more complicated maneuver sequences can provide additional savings. These three-burn sequences are called the bielliptic transfer and the biparabolic transfer.

THE BIELLIPTIC TRANSFER

A typical three-burn bielliptic transfer maneuver sequence is sketched in Figure 4.4. When a space vehicle executes a bielliptic transfer maneuver, it travels along two different elliptical coasting arcs sandwiched between three impulsive burns. The

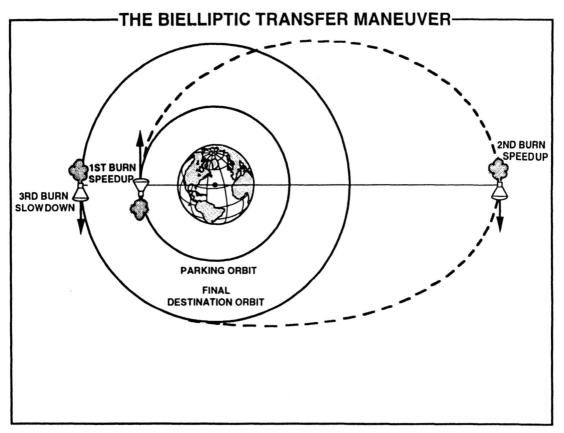

THE BIELLIPTIC TRANSFER MANEUVER

2ND BURN SPEEDUP

1ST BURN SPEEDUP

3RD BURN SLOWDOWN

PARKING ORBIT

FINAL DESTINATION ORBIT

Figure 4.4 *The famous German science writer Willy Ley once compared an astronaut executing a bielliptic transfer maneuver with a commuter who wants to travel from Los Angeles to Chicago. When the commuter calls his favorite airline, he finds that the most efficient way to get to Chicago is to go by way of New York. As Willy Ley observed, "This is quite puzzling to the commuter. Not only is New York farther away than his destination Chicago, but he goes* through *Chicago on the way!"*

first burn drives the spacecraft onto a transfer ellipse with its apogee high above the final destination orbit. The second burn, which occurs 180° away at the apogee of the first ellipse, raises the perigee of the second transfer ellipse up to the final destination altitude. The third burn, which occurs at that second perigee, in the retrograde direction, circularizes the orbit at the desired altitude.

When the German science writer Willy Ley first heard about the bielliptic transfer maneuver, he was enchanted by its strange characteristics. Soon he wrote a description in which he compared an astronaut executing such a sequence of maneuvers with a businessman who wants to fly from Los Angeles to Chicago. When he checks with the airlines, the businessman learns that the most efficient approach is to go by way of New York. "This is quite puzzling to the commuter," Willy Ley wrote with amazement. "Not only is New York farther away than his destination, Chicago, but he goes *through* Chicago on the way!"

In some cases, an even more preposterous maneuver sequence provides us with a better way to transfer from one coplanar circular orbit to another. Sometimes the best approach is to fly off to infinity along a parabolic escape trajectory and then come sailing back on a different parabola! This approach has been called the biparabolic transfer. In the real world, of course, reaching infinity along a parabolic escape trajectory would take an infinite amount of time. So, in practice, we must choose an intermediate altitude slightly lower than infinity. Moreover, the added complexity—and travel time—associated with both the bielliptical and the biparabolic maneuver sequences is intrinsically undesirable. Consequently, these complicated maneuver sequences are seldom used in the real world.

WALKING-ORBIT MANEUVERS

In the early days of the space program, aerospace engineers were delighted if they were able to launch a satellite into any stable orbit that did not immediately cause it to plunge back to Earth. Today, however, many modern satellites are targeted for very specific orbits or even specific points in space!

Each new Navstar GPS (Global Positioning System) satellite, for instance, shares its 12-hour orbital ring with three similar satellites. Moreover, all four of those satellites must be properly distributed within their orbital ring. Unfortunately, due to orbital perturbations, the four satellites occupying each ring gradually drift away from their assigned orbital slots. Consequently, every 18 months or so, on average, onboard hydrazine rockets must be fired under ground control in a two-burn sequence to reposition each GPS satellite within its orbital ring. This is accomplished by employing a two-burn impulsive sequence called a "walking-orbit" maneuver.

A typical walking-orbit maneuver sequence is shown in Figure 4.5. In this case, a retrograde burn is followed by a mul-

WALKING ORBITS:
THE ΔV COST OF VARIOUS PHASING INTERVALS

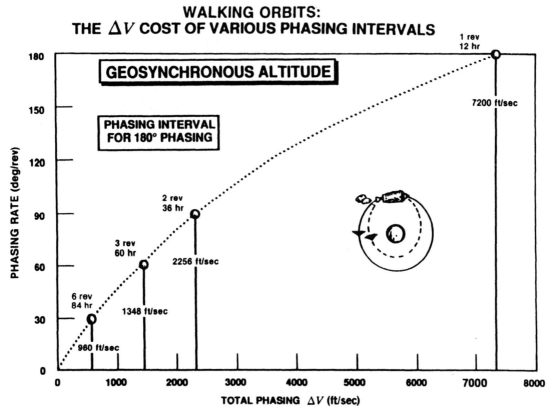

Figure 4.5 *This graph depicts the total ΔV's and the corresponding travel times required to move a satellite 180° around a geosynchronous orbit using a two-burn, walking-orbit maneuver sequence. If the mission planners are willing to burn enough propellants to generate a total ΔV of 7200 feet per second, they can move the satellite 180° in just 12 hours. But, if they have the time to execute a more leisurely maneuver sequence with two revolutions spread over a 36-hour transfer interval, only 2256 feet per second will be required. As the vertical lines in this figure indicate, only certain specific V's and certain specific travel times are possible when executing a walking-orbit maneuver under the stated conditions.*

tiorbit elliptical coasting arc of proper duration, which, in turn, is followed by a posigrade burn that puts the satellite back into its original circular orbit. Notice that when it has finished its "walk," the satellite occupies a new location with respect to the other three satellites in that orbital ring.

Notice also that the first (retrograde) burn drives the satellite into an elliptical orbit interior to the circular orbit it originally occupied. The period of this interior orbit is shorter than the period of the circular orbit. So, each time the satellite arrives at the apogee of its phasing orbit, it is at a different location with respect to the other three satellites in the original circular orbit. When it finally arrives at the desired location, a posigrade burn drives it back into its original orbit at the proper location with respect to the other three satellites.

The satellite can be made to phase in the *opposite* direction if we employ a walking orbit *external* to the original circular orbit. In this case, a posigrade burn is followed by a coasting interval of the proper duration, which, in turn, is followed by a retrograde burn to place the satellite back into its original orbit.

When a walking-orbit maneuver sequence is being executed, the satellite must travel around its elliptical phasing orbit an integral number of times before it can execute the proper recircularization burn. Consequently, only certain discrete ΔV budgets and certain discrete phasing intervals can be employed.

The graph in Figure 4.5 summarizes the permissible ΔV budgets and the corresponding coasting intervals that must be utilized in moving a commercial communication satellite 180° along the geosynchronous arc. The maneuver sequence in question is similar to the one that was executed in the 1970s when NASA's ATS-6 satellite had to be moved from its original location south of India to a completely new geosynchronous location where it could serve the continental United States.

As Figure 4.5 indicates, a geosynchronous satellite can be moved 180° along the geosynchronous arc in only 12 hours. During that 12-hour interval, it completes a single elliptical orbit around Earth. Unfortunately, we must provide it with a ΔV budget totaling 7200 feet per second. Smaller ΔV budgets totaling only 2256, 1348, and 960 feet per second can be used if longer repositioning intervals of 36, 60, and 84 hours, respectively, are acceptable. Of course, even smaller ΔV budgets are possible if longer repositioning intervals can be accepted.

ON-ORBIT RENDEZVOUS

Most of the time, when we draw a satellite orbiting Earth, we adopt an omnipotent vantage point as though we are viewing its orbit from some spot deep in outer space. In such a drawing, Earth occupies center stage.

In some cases, however—especially for rendezvous and docking studies—we are able to visualize the spatial relationships and the powered flight maneuvers more clearly if we choose to depict the motion of the moving satellite in *relative motion coordinates*. In such a depiction, the satellite will behave as though we are observing it from the vantage point onboard another moving satellite.

Fortunately, the relative motion coordinates in such a depiction can be characterized by a surprisingly simple set of equations provided the target satellite from which we are making the required observations is traveling around a circular orbit. These special equations are called the Hill-Clohessy-Wiltshire relative-motion equations.

The x, y, and z coordinate system (see Figure 4.7) rotates around the center of Earth at an angular rate that matches the angular motion of the target spacecraft. As Figure 4.6 indicates, the y axis always points radially away from the center of Earth through the center of gravity of the target spacecraft. The x axis points out the rear of the target spacecraft in the direction opposite to its current velocity vector. And the z axis protrudes out the side of the spacecraft perpendicular to its orbit plane. The roving spacecraft, which need not be in a circular orbit, is also shown in Figure 4.6. Once we have computed the three initial relative position coordinates (x_0, y_0, z_0) and the three relative-velocity components $(\dot{x}_0, \dot{y}_0, \dot{z}_0)$ of the roving spacecraft, we can use the Hill-Clohessy-Wiltshire equations listed on the right-hand side of Figure 4.7 to propagate the relative position coordinate x, y, z to any desired future time, t.

The Hill-Clohessy-Wiltshire equations can be used to construct relative-motion plots that provide powerful insights into various types of rendezvous and phasing maneuvers and their associated coasting arcs. Figure 4.8, for instance, shows how a walking-orbit maneuver would appear from the vantage point of a future astronaut who is assigned to rendezvous with a geosynchronous satellite. That courageous astronaut begins her mission in the same circular orbit as the geosynchronous satellite, but her spaceship is coasting along behind it at a range of 6000 nautical miles.

To phase forward toward the target spacecraft, she must add a rather counterintuitive *retrograde* ΔV. In the case shown in Figure 4.8, she has driven the perigee of her orbit downward 600 nautical miles. Once she is in that elliptical phasing orbit, she coasts twice around Earth along a 720° coasting arc. When

REPHASE MANEUVERS USING "WALKING" ORBITS

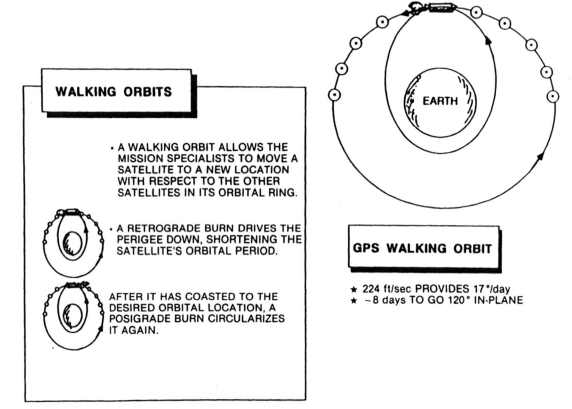

WALKING ORBITS

- A WALKING ORBIT ALLOWS THE MISSION SPECIALISTS TO MOVE A SATELLITE TO A NEW LOCATION WITH RESPECT TO THE OTHER SATELLITES IN ITS ORBITAL RING.

- A RETROGRADE BURN DRIVES THE PERIGEE DOWN, SHORTENING THE SATELLITE'S ORBITAL PERIOD.

AFTER IT HAS COASTED TO THE DESIRED ORBITAL LOCATION, A POSIGRADE BURN CIRCULARIZES IT AGAIN.

GPS WALKING ORBIT

★ 224 ft/sec PROVIDES 17°/day
★ ~8 days TO GO 120° IN-PLANE

Figure 4.6 A low-velocity, walking-orbit maneuver sequence is executed to move each GPS satellite to a new relative location within its orbital ring approximately every 18 months. For phasing in the posigrade direction, the satellite is commanded to execute an impulsive retrograde maneuver that drives it onto an elliptical *interior* phasing orbit with a shorter period. Then it coasts around that elliptical phasing orbit the requisite number of times. Each time it arrives at the apogee of the transfer ellipse, it is located at a different point with respect to the other satellites occupying that orbital ring. When it finally arrives at the desired destination, it performs a posigrade burn at apogee to recircularize its orbit. Retrograde phasing (in the opposite direction) is accomplished with a posigrade maneuver that drives the satellite onto an *external* elliptical orbit with a period that is longer than the period of its original circular orbit.

she arrives at the second apogee of that walking orbit, she must execute a *posigrade* circularization burn to match the velocity of the target spacecraft.

DEORBIT MANEUVERS

All manned space shuttles launched into low-altitude orbits must eventually execute deorbit maneuvers to get their astronauts back to Earth. In some cases, unmanned space vehicles must also reenter along purposely controlled trajecories. In the 1970s, mission planners feared that the S-II stage of the Saturn V Moon rocket, which carried the Skylab capsule into orbit, could, under certain conditions, collide with Skylab. Consequently, Skylab's trajectory specialists were assigned to study the possibility of deorbiting the S-II stage to get it safely out of the way.

The simplest approach would have been to execute a retrograde rocket burn just large enough to drive the S-II onto an elliptical orbit with perigee radius equal to the radius of Earth. If we had used that approach, the S-II stage would have traveled along a coasting arc spanning nearly 11,000 nautical miles (180°) before plunging into some remote ocean.

At the time we were assigned to carry out this reentry study, the Skylab capsule was being planned for a 262-nautical-mile circular orbit. A pure retrograde burn executed at that altitude requires a ΔV of 410 feet per second to achieve a 180° 11,000-nautical-mile reentry range. That minimum ΔV (410 feet per second) would have caused Skylab to reenter, but it would have breached Earth's atmosphere along an extremely shallow reentry trajectory, thus scattering debris fragments along an intolerably large reentry footprint. In an attempt to reduce the size of the reentry footprint, we experimented with larger ΔV's and various altitude angles during the burn that would bring the S-II stage back to Earth.

As the graph at the top of Figure 4.9 indicates, ΔV's larger than 410 feet per second would shorten the impact range—and, consequently, reduce the size of the impact footprint. Based on the contours of this graph, we selected a ΔV of 550 feet per second with corresponding reentry range of about 8000 nautical miles.

THE RELATIVE-MOTION EQUATIONS

CURRENT POSITION COORDINATES

$$x = 4\frac{\dot{x}_o}{\omega} - 6y_o \sin \omega t - \frac{2\dot{y}_o}{\omega} \cos \omega t$$
$$+ (6y_o\omega - 3\dot{x}_o)t + x_o + \frac{2\dot{y}_o}{\omega}$$

$$y = \frac{\dot{y}_o}{\omega} \sin \omega t + \left(2\frac{\dot{x}_o}{\omega} - y_o\right)\cos \omega t + 4y_o - \frac{2\dot{x}_o}{\omega}$$

$$z = \frac{\dot{z}_o}{\omega} \sin \omega t + z_o \cos \omega t$$

INITIAL RELATIVE POSITION AND VELOCITY $(x_0, y_0, z_0, \dot{x}_0, \dot{y}_0, \dot{z}_0)$

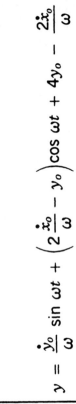

EARTH

ω

RELATIVE POSITION HISTORY FOR 2-D ORBIT PHASING

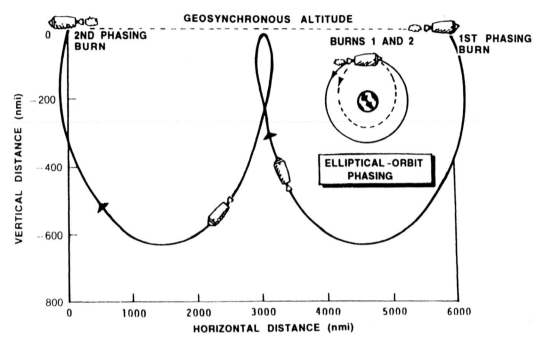

Figure 4.8 This simple relative-motion plot shows how a rendez-vousing satellite moves relative to another satellite in a co-planar circular orbit. The moving satellite is initially 6000 miles behind its target in the same circular orbit. To phase toward the target, it performs a retrograde walking-orbit maneuver that drives it onto an interior phasing orbit. After coasting twice around this elliptical phasing orbit, it passes into the vicinity of the target satellite where it executes a circularization burn.

Figure 4.7 These simple relative-motion equations can be used to calculate the instantaneous line-of-sight position of a satellite moving in the vicinity of another satellite that is traveling around a circular orbit. Notice that the y axis in the rotating coordinate system always points radially away from the center of Earth. The x axis points rearward in the direction opposite to the motion of the target satellite, and the z axis points through the side of the target spacecraft perpendicular to its orbit plane. If the initial position and velocity components of a roving satellite relative to the target satellite can be established, its x, y, z position coordinates can be propagated into the future by substituting the appropriate time, t, into these relative-motion equations.

DEORBIT MANEUVERS FOR THE SKYLAB BOOSTER

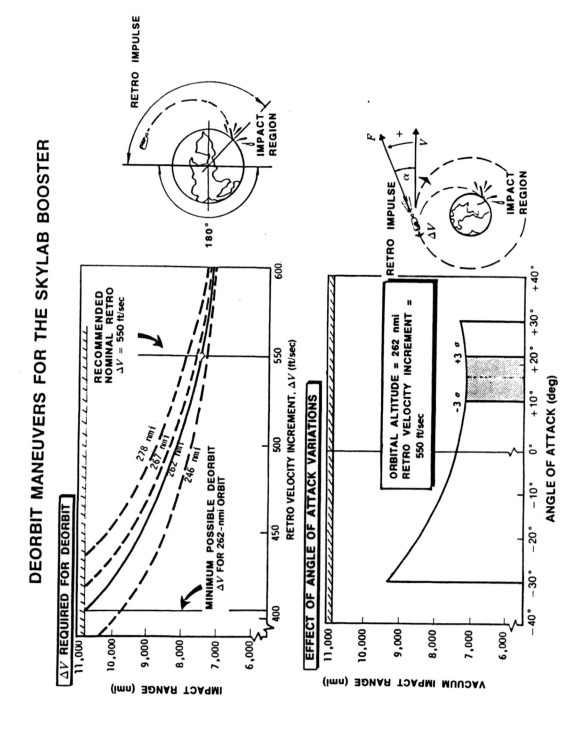

The graph at the bottom of Figure 4.9 shows how minor adjustments in the flight-path angle can further reduce the size of the impact footprint. These reductions are accomplished by tipping the vehicle into a "nose-down" altitude prior to executing its reentry burn. An angle of attack of 16° reduces the impact range by 1000 nautical miles and makes the footprint a little smaller.

Although our computer simulations indicated that deorbiting the S-II stage was a practical possibility, our managers ultimately decided to allow it to decay naturally due to friction with the atmosphere. Years later, when it finally fell back to Earth, no one, except aerospace industry professionals, paid much attention to the event. But when Skylab itself later breached the atmosphere, the public and the news media suddenly went berserk. Some clever entrepreneurs even sold Skylab insurance at statistically exorbitant rates guaranteed to compensate anyone killed by stray Skylab fragments. On the day it came back, about 200,000 Earthlings died, but not even one of their deaths was caused by Skylab!

SPACE-AGE SLINGSHOTS

When I was a high school student matriculating part time in Milburn's Pool Hall under Effie Foster's hat shop on Main Street in Springfield, Kentucky, most of my favorite teachers quietly concluded that my attempt to master that game was a complete waste of time. They did not realize that I was spending my teenage years mastering fundamental principles of physics that would later help me understand planetary swing-by maneuvers—which are now rapidly opening up the solar system to direct spaceborne exploration.

When it is executing a planetary swing-by maneuver, a space vehicle steals energy from one planet to fly by another. Figure 4.10, for example, shows how such a vehicle can swing by

Figure 4.9 The graph at the top provides the retrograde ΔV's needed to deorbit the S-II stage used in boosting the manned Skylab into orbit. After a thorough analysis, mission planners recommended an impulsive deorbit ΔV of 550 feet per second. The graph at the bottom helped in the selection of the recommended angle of attack of 16° to be used during the maneuver. This ΔV/angle of attack combination provided a footprint of reasonable size in keeping with an acceptable ΔV budget.

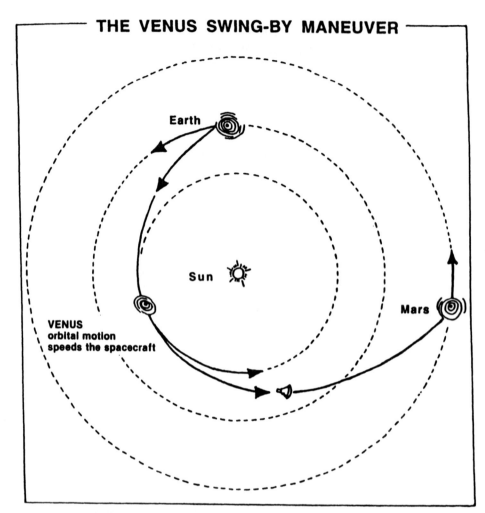

THE VENUS SWING-BY MANEUVER

Earth

Sun

VENUS
orbital motion
speeds the spacecraft

Mars

Figure 4.10 Planetary swing-by maneuvers allow a space probe to ricochet along the edge of the gravity well of one planet to reach another more distant planet. In this swing-by encounter, the spacecraft passes near the interior planet Venus to steal enough extra energy to coast out away from the Sun far enough to reach the exterior planet Mars.

Venus to pick up enough energy to reach Mars. Such a swing-by maneuver may seem like magic, but no laws of physics are violated. The spacecraft picks up energy during a swing-by maneuver (or loses it!), but the orbital energy of the planet changes by exactly the same amount. Of course, planets, are big and heavy, so they slow down or speed up by an infinitesimal amount.

One easy way to visualize what happens during the Venus swing-by depicted in Figure 4.10 is to imagine the spacecraft as a pool ball rolling into the edge of the moving whirlpool-shaped gravity well of Venus. As it enters the gently curving walls of the gravity well, it is deflected along a new trajectory with an entirely new orbital energy with respect to the Sun. The closer it comes to Venus, the deeper it dips into the Venutian gravity well, and the more it is deflected.

If the planet Venus remained stationary with respect to the Sun, a spacecraft passing near it would neither gain nor lose heliocentric energy during such an encounter. However, Venus is traveling around the sun at 78,000 miles per hour. Consequently, when the spacecraft swoops into the edge of the gravity well of Venus, it can be speeded up or slowed down in a controlled manner with the desired change in direction. Moreover, the energy change it experiences is directly analogous to what happened to the color-coded plastic balls ricocheting around the tables in Milburn's Pool Hall when the players slammed them into one another.

BALLISTIC CAPTURE MISSIONS

Space capsules that are launched from Earth into near circular orbits around the Moon usually trace out Hohmann Transfer ellipses as they travel across cis-lunar space. When this technique is used, the capsule arrives at the Moon in approximately three days. However, when it gets into the vicinity of the Moon, it must execute one or more relatively large retrograde maneuvers to slow itself down for capture by the Moon.

Fortunately, two orbital mechanics experts at NASA's Jet Propulsion Laboratory, Belbruno and Miller, have found a less costly way to handle cis-lunar missions. Their imaginative approach, which is called the *ballistic capture*, lengthens the missions duration by about a factor of 30, but, compared with the more conventional "Hohmann Transfer" approach, it saves approximately 500 feet per second.

A space capsule executing a typical ballistic capture mission to the Moon follows the rather complicated maneuver sequence sketched in Figure 4.11. Notice how the capsule passes into the vicinity of the Moon two different times during the mission. On its first encounter, it flies past the Moon while executing a trailing-edge swingby maneuver. This eases it onto the

BALLISTIC CAPTURE MISSIONS TO THE MOON

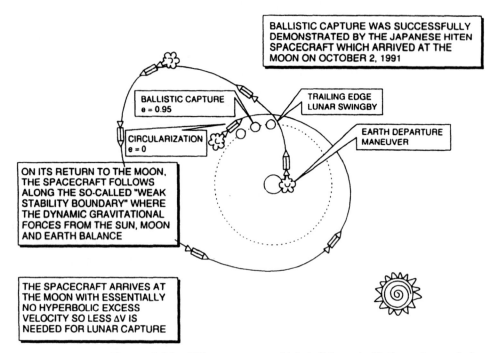

BALLISTIC CAPTURE WAS SUCCESSFULLY DEMONSTRATED BY THE JAPANESE HITEN SPACECRAFT WHICH ARRIVED AT THE MOON ON OCTOBER 2, 1991

BALLISTIC CAPTURE
e = 0.95

TRAILING EDGE LUNAR SWINGBY

CIRCULARIZATION
e = 0

EARTH DEPARTURE MANEUVER

ON ITS RETURN TO THE MOON, THE SPACECRAFT FOLLOWS ALONG THE SO-CALLED "WEAK STABILITY BOUNDARY" WHERE THE DYNAMIC GRAVITATIONAL FORCES FROM THE SUN, MOON AND EARTH BALANCE

THE SPACECRAFT ARRIVES AT THE MOON WITH ESSENTIALLY NO HYPERBOLIC EXCESS VELOCITY SO LESS ΔV IS NEEDED FOR LUNAR CAPTURE

Figure 4.11 When a space vehicle is flying a ballistic capture mission to the Moon, it first careens around the backside of the Moon where it executes a trailing-edge swingby. This precise encounter gently pulls the vehicle onto the "weak stability boundary" created by the combined gravitational fields of the Moon, Earth, and the Sun. The mission duration increases and it requires extremely precise trajectory adjustments, but a typical lunar mission can be executed, with 500 feet per second less ΔV than comparable missions that employ more conventional "Hohmann Transfer" maneuvering techniques. (*Source: Belbruno and Miller, "Sun-Perturbed Earth-to-Moon Transfers with Ballistic Capture," Journal of Guidance, Control, & Dynamics, (July–August 1993): 770–775.*

so-called "weak stability boundary" along which the gravitational forces of the Sun, the Moon, and Earth are in delicate balance.

Coasting along the weak stability boundary is conceptually akin to rollerblading along the sharp crest running along the top of a mountain chain. Small midcourse maneuvers are usually required to keep the capsule moving along this finely balanced trajectory. But, when it arrives at the Moon for its second encounter, it can achieve lunar capture with a much smaller retro ΔV.

A ballistic capture typically saves 500 feet per second compared with the more conventional Hohmann Transfer approach of reaching lunar orbit. In 1991 Belbruno and Miller demonstrated the benefits of their novel trajectory-shaping technique when they helped a team of aerospace engineers in Japan plan and execute a sequence of ballistic capture maneuvers for the Japanese Hitten spacecraft. After a little over 3 months on a complicated free-fall trajectory, the Hitten spacecraft was successfully captured in the Moon's gravity well on October 2, 1991.

5 BOOSTING A SATELLITE INTO ORBIT

Nature and nature's laws lay hid in night: God said "Let Newton be!" and all was light.—*Alexander Pope's epitaph written to commemorate the lifetime accomplishments of Sir Isaac Newton*

In 1959, I traveled on a big, silver commercial jetliner from Richmond, Kentucky, to Santa Monica, California, to my first job in the U.S. aerospace industry. When I arrived there I had only a vague idea about what a rocket was or how it worked. Gradually, however, as I came to work each day, I began to develop a speaking acquaintance with the fundamental laws of physics governing the operation of the large, multistage rockets that were hurling Soviet and U.S. satellites into space.

A rocket can be regarded as a jet engine that carries its own oxidizer. Both jets and rockets push themselves forward by expelling exhaust molecules toward the rear; however, a jet engine uses the ambient air to burn its fuel. A typical commercial jetliner burns the oxygen contained in 21 pounds of air for every pound of kerosene it consumes. A kerosene-fueled rocket burns 2.3 pounds of pure liquid oxygen for each pound of kerosene. A rocket fueled by hydrogen and oxygen operates at a completely different mixture ratio. It burns about five pounds of oxidizer per pound of hydrogen fuel.

The high-temperature gases inside the combustion chamber of a rocket create enormous pressures that push in all directions except where the exhaust nozzle allows the gases to escape. Directly opposite the exhaust nozzle is an unbalance force that hurls the rocket forward. Thus, a rocket is vaguely akin to the toy balloon, carelessly released, that traces out kamikaze spirals around your dining room.

THE ROCKET AS A MOMENTUM EXCHANGE DEVICE

A rocket can be regarded as a momentum-exchange device. Isaac Newton would have had an instant appreciation for the momentum-exchange equations listed in Figure 5.1. In this conceptual diagram, a heavy mass is attached to the rear of the vehicle with a coiled spring. When the spring tension is suddenly released, the mass is hurled rearward with a velocity, V. That event is assumed to happen in an infinitesimally brief

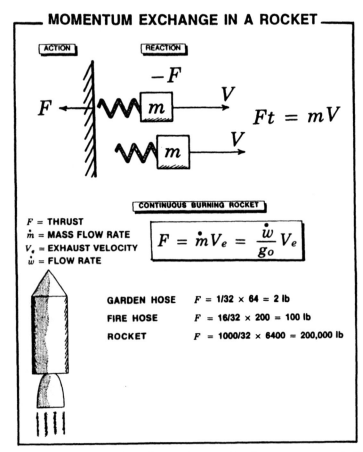

Figure 5.1 The top of the figure shows what happens to a rocket if it hurls a single large "exhaust molecule" of mass m to the right in an infinitesimally brief interval. In accordance with the momentum equation, the force times the elapsed time must equal the mass times the change in velocity. The equation at the bottom of the figure shows what happens if a similar rocket hurls out a steady stream of exhaust molecules in a continuous manner.

interval. Consequently, in accordance with the laws of New-tonian mechanics, the force times the elapsed time must equal the mass times the change in velocity:

$$Ft = mV$$

A rocket, of course, burns continuously. So the force it exerts (see Figure 5.1) can be written as

$$F = \frac{\dot{w}}{g_0} V_e$$

where $\dot{w}$ is the rate at which the rocket is burning its propellants in pounds per second, g_0 is the gravitational acceleration at the surface of Earth, and V_e is the velocity imparted to the rocket's exhaust molecules.

This equation can be used in evaluating the momentum-exchange devices highlighted at the bottom of Figure 5.1: a garden hose, a fire hose, and a single-stage rocket of modern design.

The garden hose in question expels one pound of water per second at a velocity of 64 feet per second, thus creating a reaction force of only two pounds. When you water your lawn with a garden hose, it is relatively easy to hold the hose steady against this two-pound back pressure force. The fire hose ejects 16 pounds of water per second at a velocity of 200 feet per second to create a reaction force of 100 pounds. This force is so powerful two strong firefighters are typically required to hold a firehose steady and point it toward the base of the flames.

The rocket in Figure 5.1 is hurling 1000 pounds of exhaust gases per second rearward at an effective exhaust velocity of 6400 feet per second. This creates a reaction force of 200,000 pounds. These values are fairly close to the performance parameters characterizing the J2 engines used in propelling the two upper stages of the Saturn V Moon rocket on their cislunar journey: the S-II stage and the S-IVB stage. The S-II stage was equipped with five J2 engines that burned hydrogen-oxygen to generate a total thrust of about one million pounds. The S-IVB was equipped with a single J2 engine of similar design.

ROBERT GODDARD'S CONTRIBUTIONS

The U.S. rocket pioneer Robert Goddard wrote a number of terrific technical papers summarizing the design characteristics of the liquid-fueled rockets for which he was awarded dozens of patents. In one 50-page publication on rocket technology, he included a brief discussion of a multistage rocket that could hurl a one-pound flash-powder charge from the surface of Earth on a collision course with the Moon. He went on to explore the possibility that astronomers all over the world might be able to point their telescopes toward the Moon and observe the ignition of that flash-powder charge.

From that day forward, overly diligent newspaper reporters hounded Robert Goddard unmercifully. For example, when one of his early rockets reached its design altitude of 2000 feet, one newspaper headline wryly observed: "Moon Rocket Misses Target by 239,799 and One Half Miles."

A few months later, editors of the *New York Times* wrote a searing editorial maintaining that Robert Goddard was unable to understand even the simplest principles of physics being ladled out daily in high school classrooms across the nation. In the opinion of those editorialists, Goddard's claim that a rocket could generate thrust in the vacuum of space was preposterous because, in a vacuum, the rocket would have no air molecules to push against. Actually, as Robert Goddard clearly understood, a rocket can operate perfectly well in a vacuum. As a matter of fact, it operates *better* in the vacuum than it does when its surrounded by ambient air.

Here is one convenient way for expressing the thrust generated by a rocket:

$$ F = \frac{\dot{w}}{g_o} V_e + A_e (P_e - P_o) $$

In this equation, $\dot{w}$ is the rate at which a rocket burns its fuel, g_o is the surface gravitational acceleration (32.174 ft/sec^2), V_e is the rocket's effective exhaust molecule velocity, and A_e is the cross-sectional area of the exhaust nozzle at its exit. Notice that the final multiplier in the thrust equation equals the difference between the ambient atmospheric pressure and the pressure created by the exhaust gases as they flow out the end of the rocket nozzle. From this simple equation, we can see that a rocket spewing its exhaust molecules into a vacuum always generates a higher thrust (and, in turn, a higher specific impulse) than an identical rocket that is operating in the atmosphere.

Fifty years after the editors of the *New York Times* published their scathing editorial denouncing the scientific "ignorance" of Robert Goddard, they printed a retraction stating that they had been dead wrong. Robert Goddard was, unfortunately, long since dead. But his widow appreciated that magnanimous gesture.

TYPICAL LIQUID- AND SOLID-FUELED ROCKETS

Two fundamentally different types of rockets—solid-fueled rockets and liquid-fueled rockets—are used in propelling satellites into space. Solid-fueled rockets (see Figure 5.2) were first developed by the Chinese 700 years ago in the thirteenth century A.D. These early "Chinese Fire Arrows" consisted of thin, cylindrical tubes packed with gunpowder that was ignited to achieve propulsive flight.

As Figure 5.2 indicates, most solid-fueled rockets do not burn from one end to another the way a cigarette does. More typically, a solid-fueled rocket burns radially outward from a central hole or "perforation" running along the length of its solid grain. If that central perforation is circular, the surface area being burned will increase as burning progresses, and the thrust will increase proportionally.

One way to tailor the thrust profile for a specific application is to mold the solid-fueled rocket with a star-shaped perforation. As the central protrusions formed by the star-shaped pattern gradually burn away, the hole will become more rounded, thus producing a more nearly constant thrust. By carefully choosing the exact geometry of the central perforation, it is possible to tailor the thrust profile of a solid-fueled rocket in almost any desired manner.

Large liquid-fueled rockets (see Figure 5.2) are usually designed with two separate tanks. One of these tanks contains the fuel and the other contains the oxidizer. These two fluids are pumped or forced under pressure into the combustion chamber directly above the exhaust nozzle where burning takes place.

A liquid-fueled rocket has been called "a plumber's nightmare" because it involves a complicated array of turbopumps, valves, gas generators, and various other pieces of equipment. However, despite its complexity, a liquid-fueled rocket is usually considerably more efficient than its solid-fueled counterpart. If they were not more efficient than solid-fueled rockets, liquid-fueled rockets would probably have never been perfected.

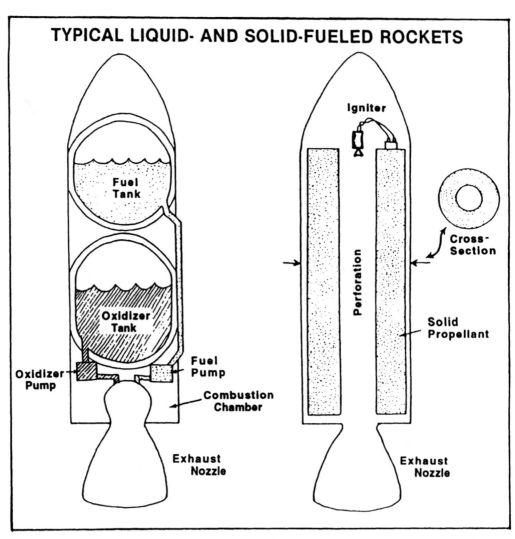

TYPICAL LIQUID- AND SOLID-FUELED ROCKETS

Igniter

Fuel Tank

Oxidizer Tank

Oxidizer Pump

Fuel Pump

Combustion Chamber

Exhaust Nozzle

Perforation

Cross-Section

Solid Propellant

Exhaust Nozzle

Figure 5.2   The first practical rockets were developed by the Chinese in the thirteenth century. The solid propellants in these "Chinese Fire Arrows" burned from one end of the cylinder to the other like a cigarette. Modern solid-fuel rockets usually burn from a central perforation radially outward toward the metal casing. Liquid-fuel rockets, which were first successfully tested in the 1930s, are usually constructed with two separate propellant tanks. One of the tanks contains the fuel; the other contains the oxidizer. These two fluids are pumped—or forced under pressure—into the combustion chamber at the rear of the rocket where burning takes place.

SPECIFIC IMPULSE

The *specific impulse* of a rocket-propellant combination is analogous to "miles per gallon" for an automobile. All other things being equal, the ΔV we can obtain from a rocket stage is directly proportional to its specific impulse. Consequently, the specific-impulse provides us with a convenient measure of a rocket's intrinsic efficiency. Once we have chosen the fuel and the oxidizer to be used, a chemical rocket's specific impulse is largely determined by the energy contained in its propellants.

The specific impulse of a rocket-propellant combination can be defined as *the number of seconds a pound of the propellant will produce a pound of thrust*. Generally speaking, rocket scientists strive for the highest specific impulse they can achieve. The specific impulse of a rocket can be computed by dividing the thrust it generates by $\dot{w}$, the rate at which it consumes its propellants:

$$I_{sp} = \frac{f}{\dot{w}}$$

This equation for specific impulse is featured in Figure 5.4 together with another useful equation for specific impulse:

$$I_{sp} = \frac{V_e}{g_o}$$

This simple relationship indicates that the specific impulse of a rocket-propellant combination is equal to the effective exhaust velocity divided by g_o, where g_o is the gravitational acceleration (32.174 ft/sec^2) at the surface of Earth.

Figure 5.3 illustrates a fanciful technique for measuring the specific impulse of a rocket. First, the rocket is mounted on top of a frictionless skateboard. Then it is attached with a string to a spring scale, which, in turn, is attached to a rigid barber pole.

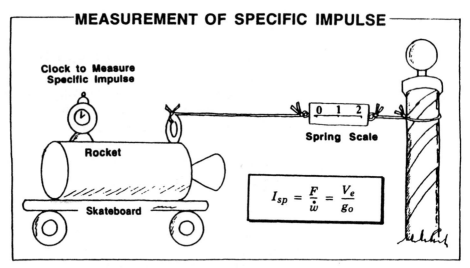

MEASUREMENT OF SPECIFIC IMPULSE

$$I_{sp} = \frac{F}{\dot{w}} = \frac{V_e}{g_o}$$

Figure 5.3 The specific impulse of a rocket-propellant combination can be defined as the number of seconds a pound of propellant can produce a pound of thrust. In this conceptual diagram, a rocket-powered skateboard is attached with a string to a barber pole that is, in turn, connected to a spring scale. To measure the specific impulse of the rocket-propellant combination, ignite the rocket and adjust its valves until it is generating one pound of thrust, as indicated by the spring scale, and then count off the number of seconds during which its one-pound-propellant load can continuously produce one pound of thrust.

To measure the specific impulse of a rocket-propellant combination, we first load the rocket with one pound of propellant (fuel and oxidizer loaded at the proper mixture ratio). Then we fire the rocket and adjust its propellant valves until the rocket is producing exactly one pound of thrust, as indicated by the spring scale. Finally, we count off the number of seconds until the rocket burns exactly one pound of propellant.

The number of seconds that elapses during the burning of one pound of propellant equals the specific impulse of that particular rocket-propellant combination. The resulting specific-impulse value, in turn, can be used in the rocket equation to calculate the velocity increment we can obtain from the rocket when it is loaded with a specific payload and a specific load of propellants.

THE ROCKET EQUATION

Figure 5.4 includes a simple derivation of the rocket equation, which is also called *Tsiolkovsky's equation*, in honor of the

Russian schoolteacher who first derived it nearly 70 years ago. Notice that the derivation hinges on the proper interpretation of Newton's second law ($F = ma$), where both the mass of the rocket, m, and its acceleration, a, are constantly changing. The end result of this step-by-step derivation is the rocket equation:

$$\Delta V = g_o I_{sp} \ln W_o / W_f$$

where ΔV is the ideal (maximum) velocity the rocket can generate, I_{sp} is the specific impulse of the rocket-propellant combination, and W_o and W_f are the ignition weight and burn-out weight respectively, at the beginning and end of the rocket burn.

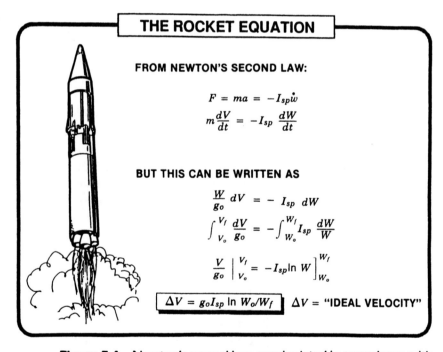

THE ROCKET EQUATION

FROM NEWTON'S SECOND LAW:

$$F = ma = -I_{sp}\dot{w}$$

$$m\frac{dV}{dt} = -I_{sp}\frac{dW}{dt}$$

BUT THIS CAN BE WRITTEN AS

$$\frac{W}{g_o} dV = -I_{sp} dW$$

$$\int_{V_o}^{V_f} \frac{dV}{g_o} = -\int_{W_o}^{W_f} I_{sp}\frac{dW}{W}$$

$$\frac{V}{g_o}\bigg|_{V_o}^{V_f} = -I_{sp}\ln W \bigg]_{W_o}^{W_f}$$

$$\Delta V = g_o I_{sp} \ln W_o/W_f \qquad \Delta V = \text{"IDEAL VELOCITY"}$$

Figure 5.4 Newton's second law, manipulated in accordance with simple relationships from integral calculus, is used in deriving the rocket equation. That famous equation, which is also called Tsiovosky's equation, is highlighted at the bottom of this figure. Tsiovosky's equation indicates that the maximum velocity we can obtain from a load of propellant is directly proportional to the specific impulse multiplied by the natural logarithm of the ratio of the weight of the rocket at ignition and the weight of the rocket at burnout.

We can use the rocket equation to calculate the ideal velocity a particular rocket can generate while burning a particular load of propellants. In the real world, of course, trajectory losses, including gravity losses, drag losses, and steering losses, must be subtracted from the ideal velocity to obtain a more realistic estimate of the actual velocity the rocket can produce.

MULTISTAGE ROCKETS

So far, all satellites orbiting Earth have been launched into space aboard multistage rockets, which have dramatically greater performance than single-stage rockets of comparable design. Mountain climbing expeditions employ a popular technique that is conceptually akin to a multistage rocket. Such an expedition might, for example, start with 100 porters and mountain climbers all of whom climb in unison one-third of the way up the mountain to their first base camp. At that level, 70 of them go back home and the remaining 30 climb another third of the way up the mountain. At that lofty level, all but six go back home. These six who remain then assault the peak.

This "staged" process is necessary because a single mountain climber cannot carry enough food and supplies to climb directly from the base of the mountain to its peak. Consequently, this complicated, but effective, "multistage" climbing technique is a far more practical approach.

The calculations in Figure 5.5 employ the rocket equation to compute the ideal velocity that can be generated by a single-stage rocket and for a two-stage rocket with exactly the same liftoff weight and the same specific impulse as the single-stage rocket.

A Typical Single-Stage Rocket

When the single-stage rocket is carrying a full load of propellants, it weighs 120,000 pounds. Its structural mass fraction is assumed to be 0.1, so, at burnout, it will weigh 12,000 pounds. A 1000-pound payload is loaded atop this 120,000-pound rocket. Consequently, at ignition, the total weight of the stack (W_o) is 121,000 pounds and the total burnout weight (W_f) is 13,000 pounds (12,000 pounds of empty casing plus 1000 pounds of payload). If we assume a specific impulse of 300 seconds, this rocket can generate a velocity increment of 21,416 feet per second. This is demonstrated by the calculations (bottom left) in Figure 5.5.

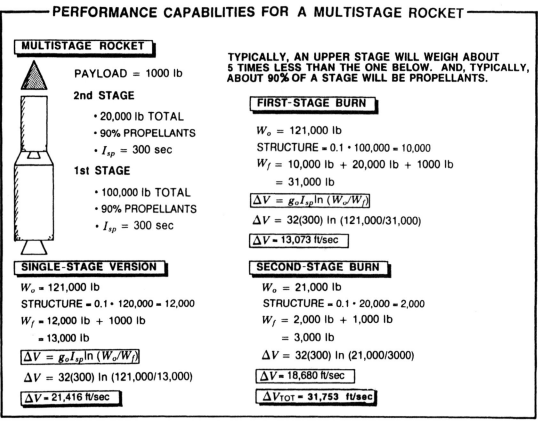

PERFORMANCE CAPABILITIES FOR A MULTISTAGE ROCKET

MULTISTAGE ROCKET

PAYLOAD = 1000 lb

2nd STAGE
- 20,000 lb TOTAL
- 90% PROPELLANTS
- I_{sp} = 300 sec

1st STAGE
- 100,000 lb TOTAL
- 90% PROPELLANTS
- I_{sp} = 300 sec

TYPICALLY, AN UPPER STAGE WILL WEIGH ABOUT 5 TIMES LESS THAN THE ONE BELOW. AND, TYPICALLY, ABOUT 90% OF A STAGE WILL BE PROPELLANTS.

FIRST-STAGE BURN

W_o = 121,000 lb

STRUCTURE = 0.1 · 100,000 = 10,000

W_f = 10,000 lb + 20,000 lb + 1000 lb

= 31,000 lb

$$\Delta V = g_o I_{sp} \ln (W_o/W_f)$$

ΔV = 32(300) ln (121,000/31,000)

ΔV = 13,073 ft/sec

SINGLE-STAGE VERSION

W_o = 121,000 lb

STRUCTURE = 0.1 · 120,000 = 12,000

W_f = 12,000 lb + 1000 lb

= 13,000 lb

$$\Delta V = g_o I_{sp} \ln (W_o/W_f)$$

ΔV = 32(300) ln (121,000/13,000)

ΔV = 21,416 ft/sec

SECOND-STAGE BURN

W_o = 21,000 lb

STRUCTURE = 0.1 · 20,000 = 2,000

W_f = 2,000 lb + 1,000 lb

= 3,000 lb

ΔV = 32(300) ln (21,000/3000)

ΔV = 18,680 ft/sec

ΔV_{TOT} = 31,753 ft/sec

Figure 5.5 A multistage rocket consists of a series of increasingly smaller rocket stages stacked one atop the other. As this simple calculation shows, a two-stage liquid rocket with a 10 percent structural mass fraction for each stage carrying a payload of 1000 pounds can generate approximately 10,000 feet per second extra ΔV compared with a single-stage rocket of similar design.

A Typical Two-Stage Rocket

Similar calculations on the right-hand side of Figure 5.5 provide us with a comparable total velocity increment for a two-stage rocket consisting of a small stage stacked atop a larger one. In this case, both stages are designed with the same structural mass fraction (0.1) as the single-stage rocket in the first case. The payload being carried into space also has the same weight of 1000 pounds. In a modern multistage rocket, the stage above typically weighs about five times as much as the stage below. This crude assumption was adopted in making the calculations running down the right-hand side of Figure 5.5, which provide correct, but nonoptimal results.

The ideal velocity generated by the first stage of this two-stage rocket is given by

$$\Delta V = g_o I_{sp} \ln W_o / W_f$$

$$= (32 \text{ ft/sec}^2)(300 \text{ sec}) \ln \frac{121{,}000 \text{ lb}}{31{,}000 \text{ lb}}$$

$$= 13{,}073 \text{ ft/sec}$$

At burnout, the first stage is discarded and then the second stage is ignited. The ideal velocity it generates is given by

$$\Delta V = g_o I_{sp} \ln W_o / W_f$$

$$= (32 \text{ ft/sec}^2)(300 \text{ sec}) \ln \frac{21{,}000 \text{ lb}}{3000 \text{ lb}}$$

$$= 18{,}680 \text{ ft/sec}$$

The sum of the two ΔV's provided by this two-stage rocket is

$$\Delta V_{\text{TOT}} = 13{,}073 \text{ ft/sec} + 18{,}680 \text{ ft/sec}$$

$$= 31{,}753 \text{ ft/sec}$$

Thus, we see that this two-stage rocket provides about 10,000 feet per second more ideal velocity than a single-stage rocket of comparable design. Even with trajectory losses, 31,753 feet per second is a large enough velocity increment to hurl the payload into a low-altitude orbit.

Trajectory Losses

The actual velocity a rocket can generate can be approximated by subtracting the appropriate trajectory losses from the ideal velocity as follows:

$$\begin{array}{cccc} \text{Ideal Velocity} & \text{Gravity Loss} & \text{Drag Loss} & \text{Steering Loss} \\ \Delta V = g_o I_{sp} \ln W_o / W_f - \displaystyle\int_{t_o}^{t_f} g \sin \gamma \, dt & - \displaystyle\int_{t_o}^{t_f} \frac{D}{m} \, dt & - \displaystyle\int_{t_o}^{t_f} \frac{F}{m} (1 - \cos \alpha) \, dt \end{array}$$

The first term in this equation is the ideal velocity that we have already calculated for the two cases. The other three terms represent the gravity loss, drag loss, and steering loss, respectively. In the real-world situations, all three of these losses are usually evaluated by numerical integration.

Gravity losses arise because part of the rocket's energy is wasted in holding it aloft and in pushing it against the relentless pull of Earth's gravity. The gravity-loss equation

$$\int_{t_o}^{t_f} g \sin \gamma \, dt$$

represents the numerical integral from the ignition point to the burnout point, where g is the local gravitational acceleration, and γ is the flight-path angle (instantaneous angle between the velocity vector and the local horizontal).

The drag loss is caused by the friction between the rocket and the ambient air. It can be expressed as

$$\int_{t_o}^{t_f} \frac{D}{m} \, dt$$

where both the drag force, D, and the mass of the rocket, m, are continuously changing. The instantaneous drag force, for example, is a strong function of the rocket's current velocity and the local density of the atmosphere.

The steering loss arises because the instantaneous thrust vector is not always parallel to the current velocity vector. This small mismatch is necessary; otherwise, we could not steer the rocket along an optimal trajectory as it flies into space. The steering loss can be evaluated from the following expression:

$$\int_{t_o}^{t_f} \frac{F}{m} \left(1 - \cos \alpha \right) dt$$

where F is the current thrust of the rocket, m is its current mass, and α is the steering angle, the angle between the thrust vector and the current velocity vector.

For a typical rocket of modern design flying into a low-altitude Earth orbit, these various losses amount to about 5000 feet per second. The velocity required to reach a low-altitude orbit is around 25,000 feet per second.

VELOCITY PROFILES

As a rocket burns away its heavy load of propellants, it begins to accelerate at an ever increasing rate. This is indicated by the solid-line and the dotted-line curves in Figure 5.6 for a single-stage rocket and a two-stage rocket, respectively. Both the single-stage and the two-stage rocket have an identical liftoff weight of 121,000 pounds and both carry a payload weighing 1000 pounds.

The burning rates for the one-stage rocket and for the first stage of the two-stage rocket are assumed to be 500 pounds per second. The second stage of the two-stage rocket is one-fifth as heavy as its first stage, so it is assumed to have a burning rate of 100 pounds per second.

The horizontal scale in Figure 5.6 represents the elapsed burning time since ignition of the first stages. The vertical scale represents the current ideal velocity. Notice that the final burnout velocity for the single-stage rocket (solid line) is 21,416 feet per second, which is the same value calculated in Figure 5.5. Notice how the rocket is building up velocity at an ever increasing rate as it approaches burnout. This increasingly rapid acceleration (change in velocity) occurs because as the rocket becomes lighter and lighter, it becomes increasingly easier to accelerate.

The dotted line in Figure 5.6 provides comparable information for a two-stage rocket with the same total liftoff weight (121,000 pounds) as the single-stage rocket. When its first stage reaches burnout, its velocity is only 13,073 feet per second. The first stage is then separated and the second stage is immediately ignited. Near burnout, both stages build up velocity at an ever increasing rate.

As we showed earlier, the total combined velocity of the two-stage rocket turns out to be 31,753 feet per second. Thus, we see that the two-stage rocket generates an extra 10,000 feet per second compared with a single-stage rocket of similar design. As the circle near the top of Figure 5.6 indicates, the single-stage rocket would need to have an extremely small structural mass fraction of only 0.0286 to provide the same ideal velocity as its single-stage counterpart. In other words, only 2.86 percent of its weight could consist of engines and structure.

These simple calculations focus our attention on the fundamental problem rocket designers face when designing and constructing a single-stage rocket that can fly useful payloads

VELOCITY PROFILES FOR 1-STAGE AND 2-STAGE ROCKETS WITH THE SAME LIFTOFF WEIGHTS

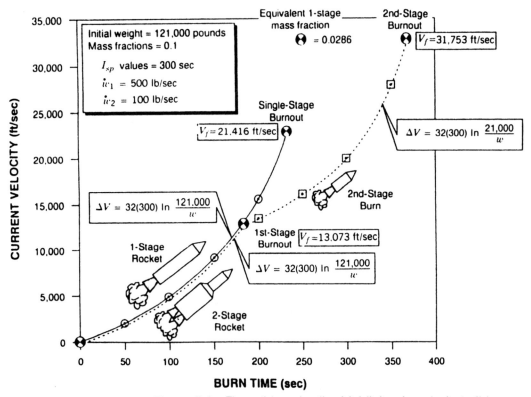

Initial weight = 121,000 pounds
Mass fractions = 0.1

I_{sp} values = 300 sec
$\dot{w}_1$ = 500 lb/sec
$\dot{w}_2$ = 100 lb/sec

Equivalent 1-stage mass fraction ⊕ = 0.0286

2nd-Stage Burnout ⊕ V_f=31,753 ft/sec

Single-Stage Burnout V_f=21,416 ft/sec ⊕

$\Delta V = 32(300) \ln \dfrac{21,000}{w}$

$\Delta V = 32(300) \ln \dfrac{121,000}{w}$

2nd-Stage Burn

1-Stage Rocket

1st-Stage Burnout V_f=13,073 ft/sec

$\Delta V = 32(300) \ln \dfrac{121,000}{w}$

2-Stage Rocket

BURN TIME (sec)

CURRENT VELOCITY (ft/sec)

Figure 5.6 The solid curving line highlights the velocity buildup rate for the single-stage rocket depicted in Figure 5.5. The dotted line depicts the comparable velocity buildup rate for the two-stage rocket in that same figure. Notice that as the rockets approach burnout, their velocity builds up at an ever-increasing rate. This increasingly rapid buildup occurs because, as the rocket burns its propellants, it becomes lighter and lighter. Consequently, it is easier to accelerate.

into space. Modern technology may eventually allow us to build a successful single-stage-to-orbit booster. Such an undertaking, however, definitely entails formidable engineering challenges.

ADDING LIGHTNESS

A quick study of the two curves in Figure 5.6 reveals that if we could somehow continue to accelerate each stage for only a few more seconds, we would be able to generate considerably higher velocities. Early rocketeers developed an insightful slogan to focus attention on a potentially fruitful engineering quest: *We need to find ways to add more lightness*. Soon they had devised a number of practical ways to add "lightness." Three specific examples are summarized in the next few paragraphs.

Balloon Tank Design

A modern rocket has been compared to an egg. In this reasonably accurate analogy, the eggshell represents the metal structure of the rocket and the yoke and the white represent its fuel and oxidizer, respectively. An eggshell makes up only about five percent of the total weight of an egg; the rest of it consists of "propellants." At first thought, it may seem difficult to imagine a rigid tank that weighs substantially less than five percent of its contents. But consider a Mickey Mouse balloon filled with pink lemonade. The balloon is so much lighter than its contents because it has no structural rigidity. It doesn't need to be rigid because its contents help support its weight. Take away the lemonade and the skin of the balloon collapses.

The designers of the Atlas rocket used a similar "balloon-tank" design to make the structure of their rocket considerably lighter than it would be otherwise. They machined the walls of the Atlas rocket so thin that it will collapse under its own weight unless either liquid or gas is kept inside to provide structural support. The resulting balloon tanks are light and efficient, but this approach is so exotic, few international rocket designers have copied the weight-saving technique pioneered at Convair in San Diego.

Common Bulkheads

Common bulkheads provide today's rocketeers with a different way to add "lightness" to a multistage rocket bound for space. Most large liquid-fueled rockets are constructed with fuel and oxidizer tanks that do not touch one another at all or else touch only at a single point. The tanks are purposely separated, in part, because a rocket's liquid fuel and the oxidizer often have grossly different temperatures. For this reason, it is prudent to avoid excessive heat transfer between the tanks.

However, introducing a common wall (or bulkhead) between the two propellant tanks can reduce substantially the total structural weight of the rocket. Its weight is lower for the same reason apartments tend to be cheaper than single-family homes. The tenants in apartments share common walls, thus conserving construction materials.

To enhance performance, we employed common bulkhead design techniques in connection with the S-II stage of the Saturn V Moon rocket. As Figure 5.7 indicates, the bulkhead protruded into the fuel tank, so the fuel and the oxidizer were separated only by a rather thin common wall. Special spray-foam insulation was used to reduce the heat transfer between the two fluids to a more acceptable level—despite the fact that the temperature differential between the two propellants was

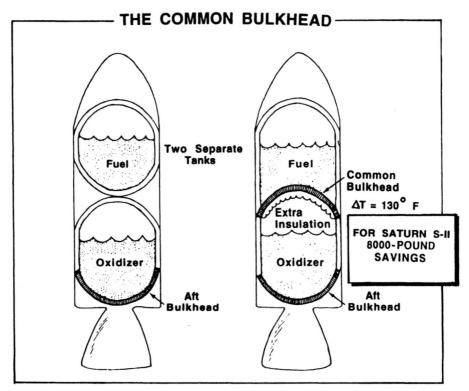

Figure 5.7 Rocket designers have figured out several ways to add lightness to their rockets to achieve enhanced performance. These diagrams show how the designers of the second stage of the Saturn V Moon rocket employed a common bulkhead to reduce weight. The common bulkhead is a shared aluminum wall between the two propellant tanks. Even after extra spray-foam insulation was added, the common bulkhead helped the designers reduce the overall stage weight by about 8000 pounds.

130° Fahrenheit. Common bulkhead design was a rather ambitious approach for its day, but it reduced the weight of the
Saturn S-II stage by 8000 pounds.

Higher-Density Propellants

Higher-density propellants provide us with another convenient method for cutting excess structural weight from a
booster rocket to be used in launching satellites into space.
Weight savings can be achieved because, all other things being
equal, higher-density propellants can be contained in tanks
that are smaller and lighter.

The propellants used in solid-fuel rockets generally have
rather high densities. Most of them range from 1.5 to 1.7 times
the density of water. The densities of liquid-fuel rocket propellants, by contrast, vary over a surprisingly broad range.

Liquid kerosene, for example, is 0.81 times as dense as water,
and the density of liquid oxygen is 1.14. Liquid hydrogen is a
high-performance rocket fuel. When it is burned in combination with liquid oxygen, it produces a specific impulse about
50 percent higher than a comparable rocket that is burning
kerosene. However, hydrogen in its liquid state is only seven
percent as dense as tap water. Consequently, a rocket fueled by
hydrogen-oxygen—such as the Saturn S-II or the Shuttle
orbiter—must be designed with huge fuel tanks to hold the
appropriate amount of liquid hydrogen.

THE HIGH COST OF ACCELERATING
UNBURNED PROPELLANTS

Almost all the energy generated by a multistage rocket bound
for space is used to accelerate propellants to be burned later by
that same rocket. If a typical satellite booster did not expend
most of its energy in this wasteful manner, it would be able to
orbit a payload weighing 30 times as much!

The calculations in the lower left-hand corner of Figure 5.8
show that if a rocket's propellants could be burned impulsively

---

Figure 5.8 The calculations on the left-hand side show that a single-
stage rocket of typical design uses about 98 percent of its energy
boosting propellants up to high velocities to be later burned. The
calculations on the right represent a similar comparison for a two-
stage rocket. For the two-stage case, the rocket uses 97 percent of its
energy in accelerating unburned propellants.

THE HIGH COST OF BOOSTING UNBURNED PROPELLANTS

SINGLE-STAGE ROCKET COMPARISON

CONVENTIONAL ROCKET

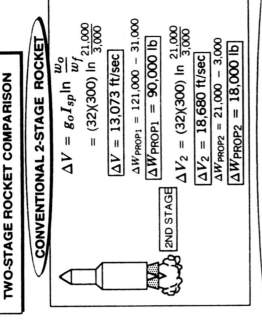

$$\Delta V = g_o I_{sp} \ln \frac{w_o}{w_f}$$

$$= (32)(300) \ln \frac{21{,}000}{3{,}000}$$

$$\boxed{\Delta V = 21{,}416 \text{ ft/sec}}$$

$$\Delta w_{PROP} = w_o - w_f$$

$$= 121{,}000 - 31{,}000$$

$$\boxed{\Delta w_{PROP} = 108{,}000 \text{ lb}}$$

SUPERBOOSTER: MOMENTUM EQUATION

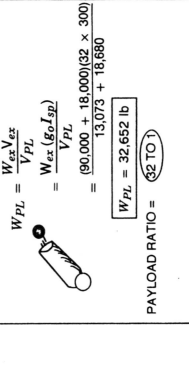

$$W_{PL} = \frac{W_{ex} V_{ex}}{V_{PL}}$$

$$= \frac{W_{ex}(g_o I_{sp})}{V_{PL}}$$

$$= \frac{(108{,}000)(32 \times 300)}{21{,}416}$$

$$\boxed{W_{PL} = 48{,}412 \text{ lb}}$$

PAYLOAD RATIO = 48 TO 1

TWO-STAGE ROCKET COMPARISON

CONVENTIONAL 2-STAGE ROCKET

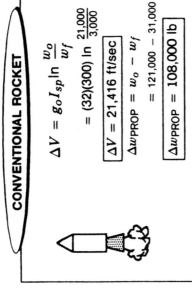

2ND STAGE

$$\Delta V = g_o I_{sp} \ln \frac{w_o}{w_f}$$

$$= (32)(300) \ln \frac{21{,}000}{3{,}000}$$

$$\boxed{\Delta V = 13{,}073 \text{ ft/sec}}$$

$$\Delta W_{PROP1} = 121{,}000 - 31{,}000$$

$$\boxed{\Delta W_{PROP1} = 90{,}000 \text{ lb}}$$

$$\Delta V_2 = (32)(300) \ln \frac{21{,}000}{3{,}000}$$

$$\boxed{\Delta V_2 = 18{,}680 \text{ ft/sec}}$$

$$\Delta W_{PROP2} = 21{,}000 - 3{,}000$$

$$\boxed{\Delta W_{PROP2} = 18{,}000 \text{ lb}}$$

SUPERBOOSTER: MOMENTUM EQUATION

$$W_{PL} = \frac{W_{ex} V_{ex}}{V_{PL}}$$

$$= \frac{W_{ex}(g_o I_{sp})}{V_{PL}}$$

$$= \frac{(90{,}000 + 18{,}000)(32 \times 300)}{13{,}073 + 18{,}680}$$

$$\boxed{W_{PL} = 32{,}652 \text{ lb}}$$

PAYLOAD RATIO = 32 TO 1

125

with perfect efficiency, 108,000 pounds of propellant burned at a specific impulse of 300 seconds could accelerate a payload weighing more than 48,000 pounds up to a velocity of 21,416 feet per second.

When we compare these calculations with the ones in the upper left-hand corner of Figure 5.8, we can conclude that 98 percent of the rocket's propulsive energy is used in accelerating unburned propellants so they will be available for burning later in the flight.

The calculations running along the right-hand side of Figure 5.8 represent a comparable case for a two-stage rocket carrying the same total propellant load of 108,000 pounds. In this case, the two-stage rocket generates a total ideal velocity of 31,753 feet per second with a 1000-pound payload. By contrast, our loss-free impulsive rocket carrying the same propellant load can handle a similar mission with a payload exceeding 32,000 pounds! The conventional two-stage rocket thus uses about 97 percent of its energy accelerating propellants to be later burned.

STAGING TECHNIQUES

When aerospace engineers design their rockets, they constantly strive to discard any extra weight as soon as it is no longer needed. Rocket stages can be stacked one atop the other (tandem staging) or they can be positioned side by side (parallel staging).

Some rockets, such as the Delta II and the Titan IV, are designed with a combination of tandem and parallel stages. The Delta II, for instance, usually employs three liquid-core stages stacked one atop the other with nine tandem strap-on solid rocket boosters arranged like a bundle of asparagus stalks around the outer periphery of its first stage.

6 TODAY'S FAMILY OF GLOBAL BOOSTERS

I have been . . . like a boy playing on the sea-shore and diverting myself now and then finding a smoother pebble or a prettier shell than ordinary whilst the great ocean of truth lay all undiscovered before me.—*Isaac Newton's humble self-evaluation (from David Brewster's* Memoirs of Newton*)*

In my first book, *The Rush Toward the Stars*, which was published in 1969, I stitched together the following series of arguments in a vain attempt to explain why reusable boosters would almost certainly turn out to be more cost-effective than the expendable boosters we were using at the time:

What makes space travel so expensive? Spectators can get a few clues as they watch the Saturn V Moon rocket beat its way off launch pad 39A and thunder into the sky. The men who built his six-million pound precision machine feel a justifiable chill of pride as they watch their brainchild hurl itself into the stratosphere, but unfortunately, the events to follow are a mechanical symphony of waste.

When the first stage reaches an altitude of 50 miles, it tips over and plummets into the atmosphere, leaving a $28 million trail of scrap metal in its wake. Six minutes later, the second stage, $30 million worth of precision machinery, forms a fireball off the coast of Africa. After two burns and a dump maneuver, the third stage whips around the trailing edge of the Moon and curls into solar orbit to become a $23 million mini planet.

If we handled routine air travel in a similar way, only kings could afford to fly. If, for example, aerial commuters had to bail out over Tokyo Bay on each flight while watching their jetliner crash into the Sea of Japan, businessmen would never be able to get their expense accounts cleared by the home office. Even at

$10,000 per ticket, the airlines would go bankrupt. It is only because the cost of the airplane can be amortized over thousands of flights that air travel can be economically feasible.

Expendable airplanes were *unintentionally* employed around the turn of the century. And as a result, air-travel costs tended to be extraordinarily expensive. On October 7, 1903, for example, Samuel Langley calmly catapulted his $50,000 airplane off the deck of a houseboat cruising along the Potomac River. It plunged straight down into the water and promptly sank. A quick calculation shows that his flight had cost about $8000 per foot!

Several other brave pioneers attempted to perfect heavier-than-air flying machines prior to the Wright brother's successful triumph. But all of them ended with a sickening splash or a tangled heap of sticks and cloth fluttering in the wind.

At first thought, a reusable space shuttle might seem to be an intrinsically cost-effective flying machine. However, the ones the Americans and the Russians have managed to construct so far, though technologically versatile and sophisticated, have not done much to pull down launch costs. Although reusability seems like a marvelous idea, reusable hardware is expensive and payloads are quite small. NASA's $1.5 billion reusable space shuttle, for instance, carries a useful payload totaling less than 1.2 percent of its liftoff weight. Servicing costs are also high. Consequently, reusability has not yet provided any reduction in the cost of launching payloads into space. Nevertheless, the space shuttle is a miracle of aerospace technology, so we will now review its capabilities and its design features in some detail.

NASA'S REUSABLE SPACE SHUTTLE

The space shuttle, whose salient features are highlighted in Figure 6.1, is a rocket when it lifts off the launch pad, a space station when it reaches Earth orbit, and a big, graceful unpowered glider when it touches down at Edwards Air Force Base or on the smooth 15,000-foot concrete runway at Cape Canaveral.

At liftoff, the fully fueled shuttle transportation system weighs approximately 4.4 million pounds. Its orbiter stage tips the scale at 150,000 pounds, and it can carry a maximum payload weighing about 50,000 pounds.

When the three main engines on the orbiter are burning, they draw liquid oxygen and liquid hydrogen from the external

NASA'S REUSABLE SPACE SHUTTLE

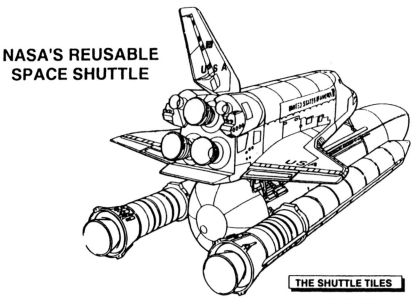

THE SHUTTLE TILES

- 30,000 TILES PER SHUTTLE
- 1 TO 5 INCHES THICK
- NO TWO EXACTLY ALIKE

SHUTTLE PERFORMANCE CHARACTERISTICS

- LIFTOFF WEIGHT = 4.4 MILLION POUNDS
- SOLID ROCKET THRUST = 2.65 MILLION POUNDS EACH
- ORBITER WEIGHT = 150,000 POUNDS
- ORBITER THRUST = 375,000 POUNDS PER ENGINE
- PAYLOAD WEIGHT = 50,000 POUNDS
- COMBUSTION TEMPERATURE = 6,000° F

THE SHUTTLE LANDING

- LANDING SPEED = 200 MILES PER HOUR
- DESCENT ANGLE = 7 TIMES COMMERCIAL LINER
- CROSSRANGE MANEUVERING = 1,200 MILES

Figure 6.1 The reusable space shuttle is part rocket, part space station, and part airplane. It takes off vertically from a special launch pad at Cape Canaveral, flies a payload into Earth orbit, and then, when it comes back, lands horizontally on a 15,000-foot runway at Edwards Air Force Base or Cape Canaveral. The space shuttle is designed for reusability, but, unfortunately, its payload totals, at most, only about 1.2 percent of its liftoff weight.

tank—which is attached to the belly of the shuttle orbiter. Each of the shuttle's main engines produce 375,000 pounds of thrust. At liftoff, this thrust is supplemented by the two strap-on solid-fueled rockets, each of which provides an additional thrust of 2.65 million pounds. The temperature of combustion for the hydrogen-oxygen mixture being burned by the shuttle orbiter is 6000° Fahrenheit—about twice the flame temperature of an acetylene cutting torch, which is designed specifically to cut through metals.

Upon reentry into Earth's atmosphere, 30,000 surface tiles—no two of which are exactly alike—protect the orbiter from excessive heating. The tiles are made from a heat-resistant material with the look and feel of Styrofoam. The 2300° Fahrenheit temperature they reach is higher than the boiling point of iron. However, the tiles are designed with amazing thermal insulating capabilities. A brave technician can place a shuttle tile into an oven, heat it to a temperature of 2300° Fahrenheit, take it out with a pair of metal tongs, and immediately hold it by the edges in his bare hand!

When the orbiter comes in for its unpowered landing, its dscent angle is 20°—about seven times the descent angle followed by a commercial jetliner. It touches down on the runway at about 200 miles per hour, landing, as television newscaster Edwin Newman once observed, "Like a butterfly with sore feet." The astronauts have only one opportunity to land the shuttle. If they miss the runway, they have no power to fly back into position for another try.

Most parts of the space shuttle are completely reusable. On a routine flight, only the large, empty external propellant tank is lost. It reaches about 97 percent of orbital velocity, and then when it is separated, it plunges back into Earth's atmosphere, overheats, and falls apart before it reaches the ground.

The solid-fuel rocket motors mounted on the two sides of the external tank are also separated during flight, but they are rigged with parachutes and they float gently back to Earth. The empty solid-fuel rockets are designed to fall into the ocean, where they are picked up by special Navy ships and towed back to Cape Canaveral to be cleaned, refurbished, and used again.

The propellants inside each of the strap-on solid-fuel rockets are partitioned into four segments. These separate segments (cylinders) are positioned one atop the other like extra-thick pancakes. Unfortunately, when the solid-fuel rockets are burning, hot spots can occur along the cracks between the various segments. Special O-rings are fastened around the rocket casing at these hazardous locations to alleviate some of the thermal loads that could damage the rocket casing. These O-rings have worked reasonably well on most shuttle missions. But, when the shuttle Challenger lifted off the pad on an unusually cold day, disaster struck when the O-ring seals were breached by the hot gases of combustion. So far, the shuttle has been launched almost 80 times with only one failure. Its overall success rate is about 98.7 percent.

THE RUSSIAN SPACE SHUTTLE

The Russian space shuttle (see Figure 6.2) bears a striking resemblance to its U.S. counterpart. Its stubby little wings, the

COMPARISONS BETWEEN THE RUSSIAN AND
U.S. SPACE SHUTTLES

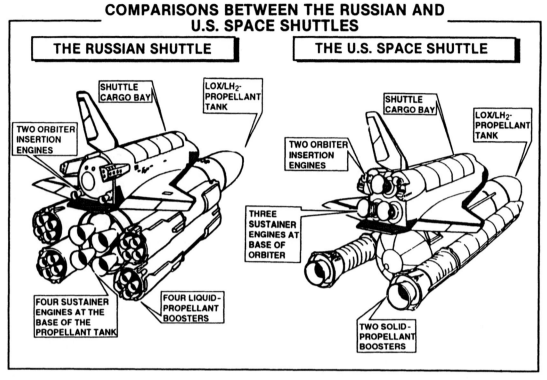

Figure 6.2 *As these sketches indicate, the Soviet space shuttle closely resembles its NASA counterpart, which became operational several years earlier. However, a few critical differences do exist. The Soviet space shuttle's core engines, which are fed by the hydrogen/oxygen center tanks, are mounted on the tanks themselves rather than on the shuttle orbiter. Moreover, the strap-on boosters nestled around the core stage are fueled with liquid kerosene and oxygen rather than solid propellants. Another important difference is that the Soviet space shuttle has been flown into space only one time, compared with more than 80 successful missions for the NASA version.*

shape of its tail, and the contours of its cylindrical fuselage all resemble the comparable parts of the U.S. shuttle. However, a few critical differences do exist between the two space shuttle designs.

The Soviet space shuttle is an all-liquid-fuel booster rocket. As Figure 6.2 indicates, it has 20 large exhaust nozzles. Notice that the four nozzles in the middle are positioned on the center core stage, not on the orbiter itself. The Russian orbiter has only two small translational thrustors for use in executing relatively small on-orbit maneuvers.

The performance of the Russian space shuttle is similar to the performance of its earlier U.S. counterpart. Its maximum advertised payload is 65,000 pounds. When the Russian orbiter comes back to Earth, it is rigged to land automatically under computer control horizontally, like a chubby little glider with only one goal in mind.

AMERICA'S EXPENDABLE BOOSTERS

The characteristics of the three workhorse boosters of the U.S. Space Program—the Delta, the Atlas, and the Titan—are discussed in the next few paragraphs. These discussions are followed by a short exploration of the key design features of the Pegasus, a new winged booster released from a Boeing L-1011 airplane at its cruising altitude prior to ignition.

The Delta Booster

Various versions of the Delta booster, which were built and marketed by McDonnell Douglas (now Boeing), have carried more than 200 satellites into space. The overall mission success for the Delta family of boosters is around 94 percent.

The Delta II Medium Launch Vehicle (see Figure 6.3) is an advanced and highly refined descendant of its simpler Thor-Delta predecessors. For most missions, the Delta II is configured as a three-stage expendable rocket with the first core stage fueled with liquid kerosene and liquid oxygen. Nine strap-on solid-fueled boosters are rimmed around the central core stage. These solid-fuel rockets are ignited prior to liftoff and they burn in parallel with the first liquid-core stage.

Various versions of the Delta booster have been used to launch low-altitude payloads, payloads bound for geosynchronous orbits, and deep-space probes. More recently, the Delta has been called upon about 30 times to launch the semisynchronous (12-hour) GPS Block II satellites. When this special need arose, the Delta production line had been shut down for many years. But, after the Challenger disaster, the line was reopened to produce an upgraded family of Delta II boosters specifically designed to launch the GPS satellites.

The Delta II can carry up to 4000 pounds into a geosynchronous transfer orbit. Under a special licensing agreement, the Japanese built their own smaller version of the Delta booster. Engineers at McDonnell Douglas are now working on a new version of the Delta, the Delta III, with larger strap-on solid-fueled rockets, a hydrogen-fueled upper stage, and other improved performance features.

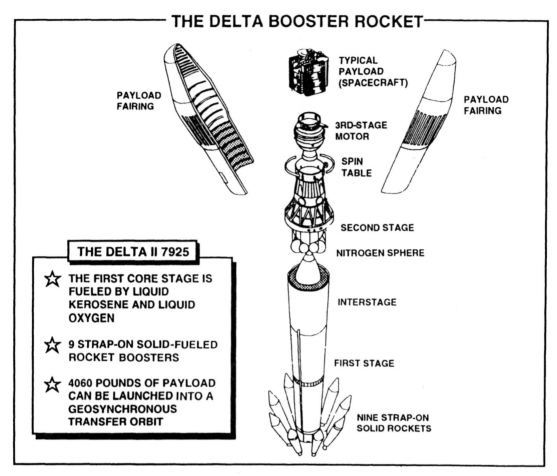

THE DELTA BOOSTER ROCKET

PAYLOAD
FAIRING

TYPICAL
PAYLOAD
(SPACECRAFT)

PAYLOAD
FAIRING

3RD-STAGE
MOTOR

SPIN
TABLE

SECOND STAGE

NITROGEN SPHERE

INTERSTAGE

FIRST STAGE

NINE STRAP-ON
SOLID ROCKETS

THE DELTA II 7925

☆ THE FIRST CORE STAGE IS
FUELED BY LIQUID
KEROSENE AND LIQUID
OXYGEN

☆ 9 STRAP-ON SOLID-FUELED
ROCKET BOOSTERS

☆ 4060 POUNDS OF PAYLOAD
CAN BE LAUNCHED INTO A
GEOSYNCHRONOUS
TRANSFER ORBIT

Figure 6.3 McDonnell Douglas designed and built the Delta II booster, which employs a core stage fueled with liquid kerosene and liquid oxygen. Nine strap-on solid boosters are rimmed around the core stage. At liftoff, the core-stage engines and the strap-on solids burn in parallel to hurl the booster and its payload off the launch pad bound for an orbit high above the stratosphere.

The Atlas Family

The Atlas booster, which is built by Lockheed-Martin in Denver, Colorado, is usually flown as a 2½- or a 3½-stage expendable rocket. Its core stage is fueled with liquid kerosene and liquid oxygen. The Atlas was the first booster used in launching the U.S. astronaut, John Glenn, into orbit aboard the compact, but efficient, Mercury capsule. Eleven Atlas boosters were later used to carry the 960-pound Block I GPS satellites into orbit.

The Atlas II can carry a 14,000-pound payload into a low-altitude circular orbit, at a 28.5° inclination (due-east launch out of Cape Canaveral). Nearly 250 Atlas launches have been executed so far with an 87 percent success rate. The Atlas booster features balloon-tank design. Its propellant tanks are fashioned from sheets of aluminum so thin the tanks are not strong enough to support their own weight. This design technique increases the vehicle's performance, but it complicates transportation and servicing of the core stage. At all times, the tanks of the Atlas must be filled with liquid propellants or pressurizing gases.

The Atlas also features stage-and-a-half design techniques. At its first staging, the heavy outer engines are separated, but no tanks are discarded at that time.

The Titan

The Titan is America's most powerful expendable booster. Its core stage, which burns unsymmetrical dimethyl hydrazine and nitrogen tetroxide, is flanked by two seven-segment, strap-on solid boosters. An early version of the Titan II was used to transport the two-man Gemini space capsules into orbit.

The uprated Titan IV can carry a 35,000-pound payload into a low-altitude circular orbit. Various versions of the Titan have been used on about 175 space missions with a success rate of approximately 93 percent.

THE WINGED PEGASUS BOOSTER

With the possible exception of the Saturn V Moon rocket, the Pegasus booster (see Figure 6.4) seems to have received more publicity for fewer successful launches than any other booster rocket in history.

The Pegasus is a three-stage all-solid-fueled booster that is released from the wing of a B-52 or the cargo bay of a modified Boeing L-1011 commercial jetliner. Its chubby little wings generate lift to help carry the booster and its payload into space. The Pegasus design teams are convinced that by releasing their all-solid-fueled booster from a subsonic aircraft at cruising altitude, they have cut its total weight by a factor of 2.

Pegasus can carry small payloads into low-altitude orbits and even smaller ones onto deep-space trajectories. It is also slated

THE PEGASUS AND THE TAURUS

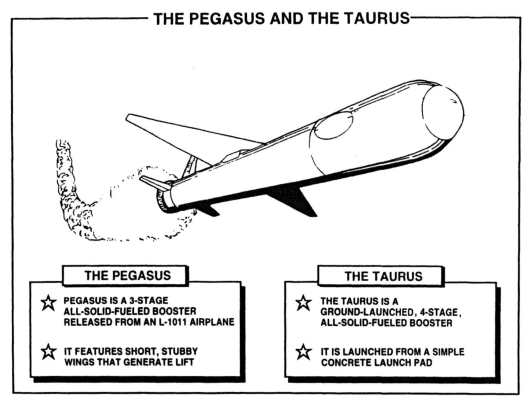

THE PEGASUS

☆ PEGASUS IS A 3-STAGE ALL-SOLID-FUELED BOOSTER RELEASED FROM AN L-1011 AIRPLANE

☆ IT FEATURES SHORT, STUBBY WINGS THAT GENERATE LIFT

THE TAURUS

☆ THE TAURUS IS A GROUND-LAUNCHED, 4-STAGE, ALL-SOLID-FUELED BOOSTER

☆ IT IS LAUNCHED FROM A SIMPLE CONCRETE LAUNCH PAD

Figure 6.4 *The Pegasus is a three-stage all-solid-fuel booster with short, stubby wings, released from a B-52 aircraft or a Boeing L-1011. Because of the extra boost provided by the airplane, the Pegasus weighs only about half as much as it otherwise would. Pegasus features Delta wings for extra lift and simple off-the-shelf design. An upgraded version, called the Taurus, is essentially a Peacekeeper upper stage with a wingless Pegasus booster stacked on top. Taurus is designed so it can be launched from a simple concrete pad.*

to deliver the 85-pound Globalstar mobile communication satellites eight at a time into low-altitude orbits. The maximum payload capability of the Pegasus is a bit less than 1,000 pounds.

Pegasus, which experienced two launch failures in its first five missions, is named after the winged horse from Greek mythology. Whenever possible, its designers have employed standard off-the-shelf components in constructing their chubby little flying machine. Costs per mission are relatively low, but in terms of dollars per pound of payload sent into orbit, the Pegasus booster provides a relatively expensive ride into space.

Orbital Sciences Corporation engineers, who designed and constructed the Pegasus, have used their expertise and experience in developing a new all-solid-fuel booster called the Taurus. Taurus, which can carry a 3500-pound payload into a 28.5° 100-nautical-mile orbit, sports a first stage borrowed from the Peacekeeper ICBM. A wingless, 3-stage Pegasus is stacked atop the Peacekeeper stage to form a four-stage all-solid-fuel rocket.

THE EUROPEAN ARIANE

The European Ariane (see Figure 6.5) has turned out to be one of the world's most successful booster rockets. In its usual configuration, the Ariane is a three-stage expendable booster

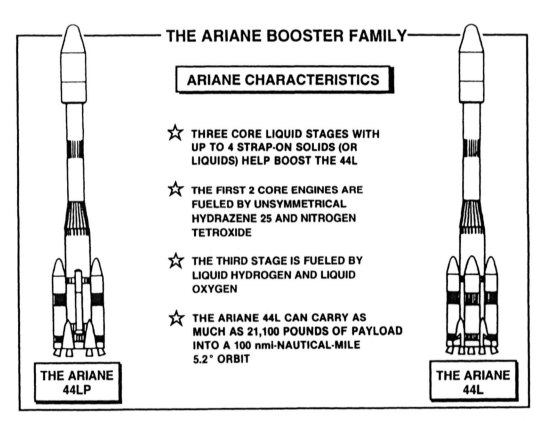

THE ARIANE BOOSTER FAMILY

ARIANE CHARACTERISTICS

☆ THREE CORE LIQUID STAGES WITH UP TO 4 STRAP-ON SOLIDS (OR LIQUIDS) HELP BOOST THE 44L

☆ THE FIRST 2 CORE ENGINES ARE FUELED BY UNSYMMETRICAL HYDRAZENE 25 AND NITROGEN TETROXIDE

☆ THE THIRD STAGE IS FUELED BY LIQUID HYDROGEN AND LIQUID OXYGEN

☆ THE ARIANE 44L CAN CARRY AS MUCH AS 21,100 POUNDS OF PAYLOAD INTO A 100 nmI-NAUTICAL-MILE 5.2° ORBIT

THE ARIANE 44LP

THE ARIANE 44L

Figure 6.5 The European Ariane is usually a three-stage expendable rocket equipped with two strap-on solid-fuel or liquid-fuel boosters. It can be used to carry payloads into low-altitude, medium-altitude, and high-altitude Earth orbits. At the peak of its popularity, the Ariane captured 50 percent of the Western world's commercial launch vehicle market.

whose core stages burn kerosene-oxygen propellants. It can be rigged with strap-on solid-fueled or liquid-fueled boosters rimmed around the first core stage. The Ariane booster is financed by a consortium of European nations with France and Germany as the dominating partners. At the peak of its popularity, various versions of the Ariane captured more than 50 percent of the Western world's commercial launch vehicle market.

On a mission bound for a 100-nautical-mile circular orbit at a 5.2° inclination, the Ariane 44L can carry a hefty payload weighing 21,000 pounds. Ariane's engineers are working on an advanced version called the Ariane V. It is equipped with a high-performance liquid oxygen-hydrogen upper stage. When it is fully operational and uprated, the Ariane V will be capable of carrying 16,000 pounds of payload into geosynchronous orbit. Unfortunately, on its first mission, the Ariane V blew up.

SOVIET BOOSTERS ON PARADE

Figure 6.6 highlights the large family of capable boosters developed, tested, and flown by the professional engineers and rocket scientists in the former Soviet Union. Their rockets range from the relatively small SL-8 with a low-altitude payload capability of 3740 pounds to the enormous SL-17 (Energiya booster), which can carry 220,000 pounds to a similar low-altitude destination.

Soviet rocketeers have consistently concentrated on manufacturing large numbers of identical boosters with hardly any design changes. Their rockets have carried enormous numbers of satellites into low-, medium-, and high-altitude flight regimes and toward the distant planets including, in particular, Venus and Mars.

Rocket scientists in the former Soviet Union tend to favor liquid-fuel engines equipped with large numbers of clustered rockets and exhaust nozzles. Some Russian boosters are equipped with as many as 20 exhaust nozzles mounted side by side.

Russian rockets fly into low-altitude orbits for military reconnaissance, medium-altitude destinations for navigation and real-time commercial communications, and geosynchronous orbits for some weather observation and communication missions. In general, however, the Russian aerospace engineers tend to favor 12-hour Molniya communication satellites posi-

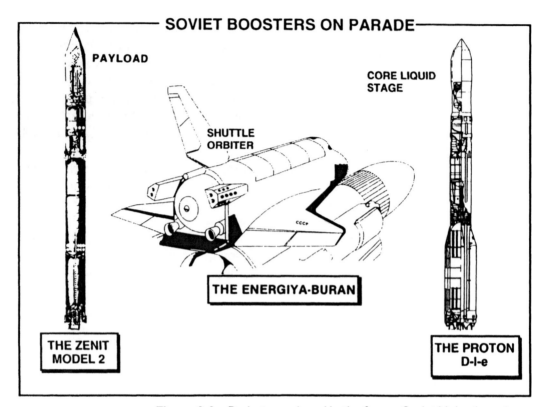

Figure 6.6 Rockets produced in the former Soviet Union have been used to transport huge numbers of satellites into the low-, medium-, and high-altitude flight regimes—and on deep-space missions bound for the distant planets. The maximum low-altitude payloads carried aloft by Soviet boosters range from about 4000 to 240,000 pounds. Despite their limited access to some forms of high-technology, the Soviets have achieved launch success rates ranging from 88 to 100 percent.

tioned in 63.4° elliptical orbits to provide abundant coverage for their high-latitude territories. In their first 2200 flights, the various families of Soviet boosters managed to achieve success rates ranging from 88 to 100 percent.

The largest Soviet booster, the Energiya, has a thrust of 8.8 million pounds. Flown in its cargo mode, it can loft 220,000 pounds into a low-altitude Earth orbit. Its central core stage, which burns liquid hydrogen mixed with liquid oxygen, is surrounded by four strap-on rockets burning liquid kerosene and liquid oxygen. The Energiya first flew into orbit in May 1987. In the meantime, it has completed only one other flight.

THE JAPANESE H-2

The Japanese H-2, worthy successor to their simpler H-1, features a core stage powered by two liquid hydrogen-oxygen engines flanked by two large strap-on solids.

The H-2 is designed to carry low-altitude Earth observation satellites and communication satellites toward specific slots along the geosynchronous arc. It will also transport the small Japanese "Hope" space plane on rendezvous missions bound for the International Space Station.

The Japanese H-2 is designed to carry 4400 pounds into a geosynchronous orbit or 8800 pounds onto a geosynchronous transfer orbit. All of its first 20 launches were executed successfully. The major problem with the H-2 is excessive cost. It is a reliable and efficient booster, but its costs are excessively high in comparison with other world-class boosters.

THE CHINESE LONG MARCH

In 1987, at the Paris Air Show, Chinese officials offered their Long March booster to other countries and consortiums with dedicated launches costing only $30 million each. The smallest versions of the Chinese Long March share the performance range of the Pegasus. The largest ones are roughly competitive with the much more capable Atlas Centaur.

In their first 25 flights, various Long March boosters delivered 23 payloads into orbit for a mission success rate of approximately 92 percent. A Long March booster weighing 444,000 pounds can carry a 3000-pound payload onto a geosynchronous transfer orbit.

ADVANCED BOOSTER CONCEPTS EMERGING FROM THE DRAWING BOARDS

Several new booster rockets are presently under development at various locales around the world. These include the X-33 reusable space plane, the Ariane V and the Delta III.

The X-33 Reusable Space Plane

As a replacement for the space shuttle for the twenty-first century and beyond, three different companies—Lockheed

Martin, McDonnell Douglas, and Rockwell International—proposed three fundamentally different versions of NASA's reusable space plane.

The X-33 is the predecessor to a single-stage-to-orbit reusable vehicle fueled by liquid oxygen and liquid hydrogen. If successful, it will carry relatively large payloads into low-altitude orbits at one-tenth the cost per pound of today's least expensive boosters. All three of the competitors proposed automated vehicles capable of executing all boost and reentry maneuvers without assistance from a human pilot.

The two versions concocted by Lockheed Martin and Rockwell International featured vertical takeoff and horizontal landing—in maneuver sequences similar to those executed by today's space shuttle. The version proposed by McDonnell Douglas called for vertical takeoff and vertical landing. That imaginative version was patterned after the McDonnell Douglas Delta Clipper, which has been flown in subscale flight tests several times at the White Sands Missile Base in New Mexico.

Lockheed's famous "skunk works" came up with the winning design. It features an advanced aerospike engine in which the propellants are mixed and burned adjacent to a special external "nozzle." Lockheed-Martin's engineers are targeting a payload delivery cost of only about $1000 per pound.

The Ariane V

The $6.7 billion Ariane V was originally designed to boost Europe's now defunct Hermes Space Plane on manned rendezvous flights bound for the International Space Station. The Ariane V's first launch ended in failure, but when it is fully operational, it will be able to deliver two satellites into geosynchronous orbits with a combined total weight of 13,000 pounds. If proposed upgrades are successfully implemented, the Ariane V will be able to deliver two satellites with a combined total weight of 16,000 pounds along the geosynchronous arc.

The Delta III

The Delta III, a grossly improved version of the Delta II, is being designed and constructed by the engineers and technicians at McDonnell Douglas (now a part of Boeing). As Figure 6.7 indicates, the Delta III will feature a new high-performance hydrogen-oxygen upper stage together with nine larger and

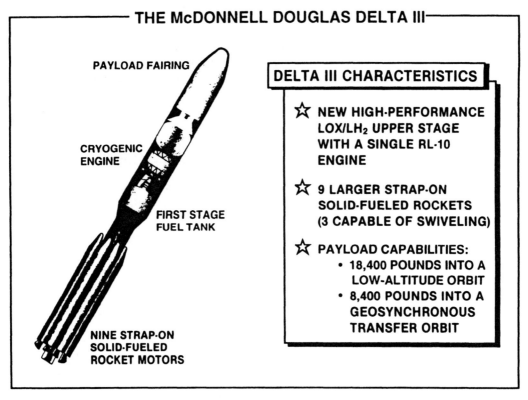

THE McDONNELL DOUGLAS DELTA III

PAYLOAD FAIRING

CRYOGENIC ENGINE

FIRST STAGE FUEL TANK

NINE STRAP-ON SOLID-FUELED ROCKET MOTORS

DELTA III CHARACTERISTICS

☆ NEW HIGH-PERFORMANCE LOX/LH₂ UPPER STAGE WITH A SINGLE RL-10 ENGINE

☆ 9 LARGER STRAP-ON SOLID-FUELED ROCKETS (3 CAPABLE OF SWIVELING)

☆ PAYLOAD CAPABILITIES:
  • 18,400 POUNDS INTO A LOW-ALTITUDE ORBIT
  • 8,400 POUNDS INTO A GEOSYNCHRONOUS TRANSFER ORBIT

Figure 6.7 The Delta III booster, which is being designed and produced by McDonnell Douglas (now Boeing), is a grossly upgraded version of their Delta II. It features a new hydrogen-oxygen upper stage and nine uprated strap-on solid rockets, three of which are capable of swiveling to achieve real-time controllability and trajectory shaping. Performance specifications call for a booster that can carry 8400 pounds onto a geosynchronous transfer orbit or at least 18,000 pounds into a low-altitude orbit.

more efficient strap-on solids—three of which will be capable of swiveling to provide thrust vector control for the rocket. The payload capability of the new booster is slated to be

· 18,400 pounds into a low-altitude circular orbit
· 8400 pounds onto a geosynchronous transfer orbit

High-level executives at Hughes convinced McDonnell Douglas to design and build the Delta III by agreeing to book 10 firm launches, sight unseen. The first launch of the new booster is scheduled for 1998. On that first mission, it will carry an unspecified geosynchronous satellite on an unspecified mission for Hughes.

NOVEL CONCEPTS FOR THE FUTURE

In this final section, we will discuss three imaginative launch techniques that have been proposed over the past several years for lowering payload delivery costs to a substantial degree: the big, dumb booster, the asparagus-stalk booster, and the mag-lifter concept.

Arthur Schnitt's Big, Dumb Booster

The earliest of the three unusual booster concepts was proposed in the 1970s by Arthur Schnitt, a design engineer at Aerospace Corporation in El Segundo, California. Schnitt's engineering assignment called for an effective new booster that could carry large and heavy payloads into Earth orbit at affordable rates. He started out, as most of his predecessors had, by trying to develop a minimum-cost, low-weight booster in keeping with the conventional wisdom swirling through the aerospace industry.

However, as he attempted to develop and refine his new booster, he began to suspect that minimum weight was intrinsically incompatible with minimum cost. Eventually, he concluded that a big, dumb booster consisting of a stack of stages with relatively poor mass fractions would probably provide the cheapest way to launch heavy payloads into space. Schnitt also concluded that the lower stages should be big, heavy, and inefficient, but also cheap and simple to produce. His upper stages were to be built at a slightly higher cost per pound with somewhat smaller mass fractions.

As Table 6.1 indicates, Arthur Schnitt's big, dumb booster was to be equipped with simple, pressure-fed engines. All stages would be fueled with inexpensive liquid propellants, and all would be large, cheap, and expendable.

The propellants he selected were nitrogen tetroxide and unsymmetrical dimethyl hydrazine. Simple pressure-fed engines were chosen rather than higher-performance, but more costly, engines fed by turbopumps. Schnitt also decided to avoid expensive and complicated regenerative cooled engines in favor of engine nozzles rigged with ICBM heat-shield linings. Liquid injection control was found to be cheaper than the more complicated and delicate swiveling engines. Even aluminum structural elements were discarded to help hold down costs. Schnitt's design documents called for the use of structurally rigid monocoque tanks made from the structural steel used in constructing submarine hulls.

TABLE 6.1   Arthur Schnitt's Big, Dumb Booster Concept

Design Characteristic	Sleek and Smart Booster	Big, Dumb Booster
Overall design criteria	Minimum-weight/maximum-performance design	Minimum-cost design
Engines	Liquid rockets (with turbopumps) and strap-on solids (optional)	Liquid pressure-fed rockets
Propellants	Liquid oxygen, kerosene, UDMH, liquid hydrogen	Storable liquid propellants: nitrogen tetroxide and UDMH
Engine cooling	Regeneratively cooled	"Heat shield" lining for the engines
Steering techniques	Swiveling engines	Fixed engines with liquid injection control
Tankage	Chemical and machine milling. Balloon tanks (optional)	Structurally rigid monocoque tanks
Structural materials	High-strength aluminum	Moderate-strength structural steel from submarine hulls
Propellant temperatures	Some cryogenic propellants	Ambient-room-temperature propellants
Mass fractions	Extremely low or low for all stages	Higher for lower stages, lower for upper stages
Manufacturing techniques	State-of-the-art (clean room construction)	Commercial production techniques
Cost per pound in orbit	High	Low

The big, dumb booster was intentionally designed to be heavier (but cheaper!) than most of the spaceborne rockets proposed before or since. It was never financed or constructed, but the privately developed Conestoga booster employed a similar engineering philosophy and some of the same design techniques.

Ed Keith's Asparagus-Stalk Booster

Ed Keith, a respected and innovative engineer who works at the Microcosm Corporation in Torrance, California, has proposed a so-called "asparagus-stalk booster" intended to hold

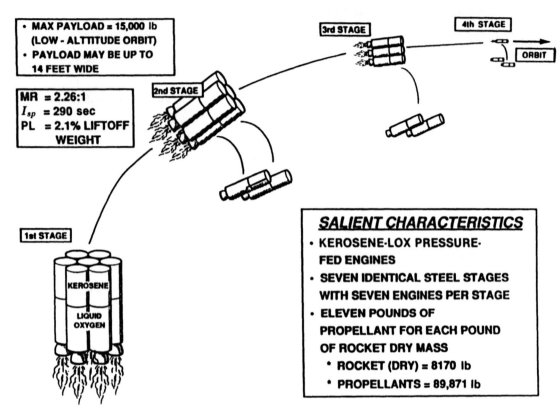

- MAX PAYLOAD = 15,000 lb
  (LOW - ALTTITUDE ORBIT)
- PAYLOAD MAY BE UP TO
  14 FEET WIDE

MR = 2.26:1
I_{sp} = 290 sec
PL = 2.1% LIFTOFF
      WEIGHT

1st STAGE
KEROSENE
LIQUID
OXYGEN

2nd STAGE

3rd STAGE

4th STAGE

ORBIT

SALIENT CHARACTERISTICS

- KEROSENE-LOX PRESSURE-
  FED ENGINES
- SEVEN IDENTICAL STEEL STAGES
  WITH SEVEN ENGINES PER STAGE
- ELEVEN POUNDS OF
  PROPELLANT FOR EACH POUND
  OF ROCKET DRY MASS
  - ROCKET (DRY) = 8170 lb
  - PROPELLANTS = 89,871 lb

Figure 6.8 This asparagus-stalk booster, which was masterminded by aerospace engineer Ed Keith, features seven cylindrical rocket stages stacked side by side, fueled by inexpensive liquid kerosene and liquid oxygen propellants. As the rocket flies, some of its propellants are automatically pumped from one stage to another. This allows some of the stages to burn all the way into orbit while others are being dropped off in pairs along the way.

Figure 6.9 The initial design for the maglifter booster employs a magnetically levitated carrier sled accelerated up to a velocity of 600 miles per hour using an electromagnet catapult sloping up the side of a 14,000-foot mountain. The maglifter is designed to handle 5000-pound payloads with peak accelerations of only 3 g's. With usage rates of two or three launches per week, this unique single-stage-to-orbit vehicle may be able to carry payloads into orbit at one-tenth the price being paid for rides aboard today's chemically fueled launch vehicles.

THE "MAGLIFTER" BOOSTER CONCEPT

MAGLIFTER GEOMETRY

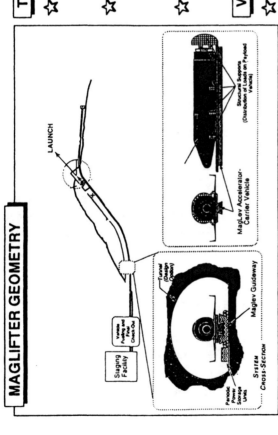

THE MAGLIFTER CONCEPT

☆ MAGNETICALLY LEVITATED "CARRIER SLED" USING SUPERCONDUCTING ELECTROMAGNETIC CATAPULT

☆ SINGLE-STAGE-TO-ORBIT VEHICLE RELEASED AT A 14,000-FOOT MOUNTAINTOP ELEVATION AT 600 MILES PER HOUR

☆ ESTIMATED DEVELOPMENT COST = $2 BILLION

VITAL STATISTICS

☆ VEHICLE WEIGHT = 500,000 POUNDS

☆ SLED WEIGHT = 200,000 POUNDS

☆ PAYLOAD WEIGHT = 5,000 POUNDS

☆ PEAK ACCLERATION = 3 g's

☆ ANTICIPATED LAUNCH RATE = 2 TO 3 LAUNCHES PER WEEK (WITH 3 REUSABLE SINGLE-STAGE-TO-ORBIT VEHICLES)

MAJOR PROPONENTS: NASA'S ADVANCED CONCEPTS OFFICE, THE AIR FORCE PHILLIPS LABORATORY

FOLLOW-ON CONCEPT: SECOND MAGLIFTER INSTALLATION IN ECUADOR AT 20,000-FOOT ALTITUDE DEVOTED TO NEAR-EQUATORIAL LAUNCHES

down space transportation costs. His concept calls for a simple, but inexpensive, multistage rocket designed with mass production in mind.

As Figure 6.8 indicates, the asparagus-stalk booster consists of seven cylindrical rockets mated side by side. When it lifts off the launch pad, all seven stages are burning in parallel. At each of the three staging points, two of the rockets are discarded.

Like Arthur Schnitt's big, dumb booster, Ed Keith's brainchild features dull-as-dishwater design. But his short, stubby multistage rocket does have one exotic design feature. As it flies toward its destination in space, its kerosene and oxygen propellants are pumped from some of the parallel stages into others. This is necessary because some of the stages burn for longer intervals. This approach may seem a bit strange, but as Ed Keith points out, airplane pilots frequently pump fuel from one tank to another during flight in essentially the same way.

Each stage of the asparagus-stalk booster is designed with seven identical stages nestled together in a simple parallel configuration. Its 15,000-pound payload makes up 2.1 percent of the booster's liftoff weight. The seven parallel rockets carry 11 pounds of propellant for each pound of structural weight.

The Maglifter Concept

Figure 6.9 illustrates how the maglifter, a magnetically levitated booster may be used to carry small payloads into orbit at economically affordable rates. The maglifters carrier sled is accelerated by superconducting electromagnets that levitate and catapult both sled and payload up to a subsonic velocity along the sloping walls of a tall mountain. Once it is released from the carrier sled, the upper-stage rocket is ignited to drive the payload into its final destination orbit.

The initial design for the maglifter calls for an electromagnetic catapult consisting of a long, straight ramp attached to the side of a 14,000-foot mountain. At the top of the mountain, the payload and its upper stage are released at a velocity of 600 miles per hour. Because the electromagnetic catapult accelerates the payload to a relatively low velocity over such a long range, the peak acceleration amounts to only about 3 g's—roughly the same as today's reusable space shuttle.

The Maglifter concept is being masterminded by NASA's Advanced Concept Office and the Phillips Laboratory, which is financed by the Air Force. Proponents are anticipating a peak mission rate of two or three launches per week using reusable single-stage-to-orbit vehicles, which are being designed to fly back to the ground and land horizontally on special landing pads. A follow-on concept with a larger maglifter may later be installed in Ecuador with a 20,000-foot release altitude. This larger version with its superior location would be especially efficient when used for launching payloads bound for the geosynchronous equatorial orbits.

7 ENHANCING THE PERFORMANCE OF BOOSTER ROCKETS

Where the statue stood of Newton with his prism and silent face: The marble index of a mind forever voyaging through strange seas of thought alone. —*William Wordsworth* (The Prelude, *1850)*

In 1961, when President John F. Kennedy delivered his famous speech asking Congress to help him conquer the Moon, America had not yet orbited a single astronaut. Yet nine years later, largely due to the efforts of one single individual armed with one simple, creative solution, Neil Armstrong and Edwin Aldrin—two lighthearted gazelles—were frolicking across the lunar surface.

At the time President Kennedy stood before Congress, Werner von Braun and his colleagues at NASA were considering two approaches for conquering the Moon. In their first concept (see Figure 7.1), they intended to build a giant 12-million pound booster rocket that would carry the astronauts directly to the Moon and back, dropping off increasingly smaller stages along the way.

That 12-million-pound booster was about 100 times bigger than any booster U.S. rocketeers had managed to construct up until that time. So, just in case it was impossible to build such a massive rocket on schedule, the von Braun team devised a second Moon-mission concept, which is also illustrated in Figure 7.1. In this case, two booster rockets weighing six million pounds each would fly into Earth orbit, rendezvous with one another, join their payloads, and then fly the mission using the same maneuvers as in the first concept.

VON BRAUN'S TWO METHODS FOR CONQUERING THE MOON

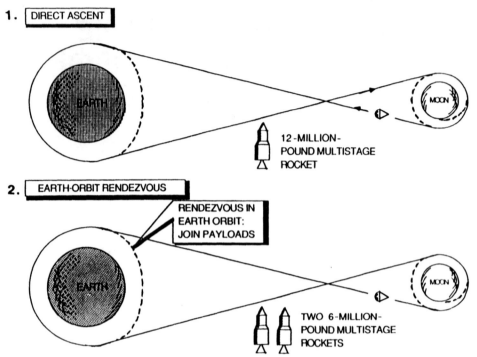

Figure 7.1 In 1961, when President Kennedy asked Congress to approve Project Apollo, Werner von Braun and his colleagues were preparing two candidate approaches for conquering the Moon. In their first concept, they intended to use a single 12-million-pound Nova booster that would fly directly to and from the Moon, dropping off increasingly smaller stages along the way. In their second concept they planned to employ two 6-million-pound boosters, rendezvous them in Earth orbit, join their payloads, and then fly the mission the same way as in the first concept.

LUNAR-ORBIT RENDEZVOUS

At first, the two concepts shown in Figure 7.1 appeared to be the only realistic approaches for conquering the Moon. But then one clever individual armed with one simple, creative solution figured out how to carry out the mission in a considerably more efficient way. His name was John Houbolt, a NASA engineer. Houbolt argued that instead of building one 12-million-pound booster for a direct flight, or two half-size boosters rigged to handle Earth-orbit rendezvous, we should, instead, carry out the rendezvous in orbit about the Moon.

But what difference would that make? John Houbolt went one to explain that if the astronauts executed their rendezvous maneuvers in orbit around the Moon, they could detach a small lifeboat-type spacecraft fly down to the surface of the Moon, and then fly an even smaller spacecraft back up again to rendezvous with the much heavier spacecraft circling overhead. That craft would include the propellants needed to break away from Moon's gravity and the heavy heat shield needed to help the astronauts survive their fiery journey through Earth's atmosphere.

As Figure 7.2 indicates, John Houbolt's clever architecture allowed the mission planners to cut the liftoff weight of the booster rocket in half. In other words, they could carry out the mission with *one single six-million-pound booster!*

Houbolt encountered stiff resistance at NASA, but his engineering studies convinced him and his colleagues that the lunar-orbit rendezvous approach would be cheaper, faster, and safer than either of the other two techniques. After a series of contentious meetings with NASA representatives, Werner von Braun and his counterparts at the other NASA centers ended up embracing John Houbolt's innovative lunar-orbit rendezvous concept. Ultimately, all of the Apollo missions that were flown to the Moon employed lunar-orbit rendezvous.

In the rest of this chapter, we will discuss various other techniques for enhancing the performance of large booster rockets. Generally speaking, these breakthroughs were accomplished with shear human intelligence—without any modifications to the existing hardware. These imaginative techniques will include propellant utilization sysstems, the programmed mixture ratio scheme, optimal fuel biasing, and optimal trajectory shaping.

PROPELLANT UTILIZATION SYSTEMS

A large liquid rocket, such as the S-II stage of the Saturn V Moon rocket, usually includes two separate tanks—one for the fuel and the other is for the oxidizer. These two fluids are pumped or forced under pressure into the combustion chamber immediately above the exhaust nozzle, where propellant burning takes place.

If we would load 1000 rockets with the required quantities of fuel and oxidizer, and then fly them to their destination

JOHN HOUBOLT'S LUNAR-ORBIT RENDEZVOUS

3. | LUNAR-ORBIT RENDEZVOUS |

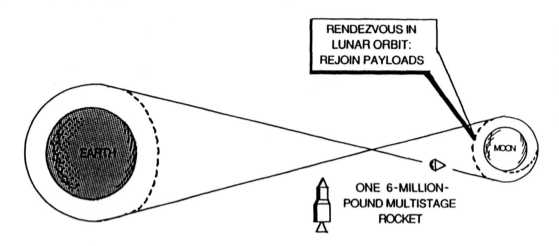

• BY SENDING A SMALL "LIFEBOAT" DOWN TO THE LUNAR SURFACE AND AN EVEN SMALLER ONE BACK UP AGAIN, THE REQUIRED BOOSTER BECAME MUCH SMALLER.

• LUNAR-ORBIT RENDEZVOUS ALLOWED NASA TO REDUCE THE LIFTOFF WEIGHT BY 6 MILLION POUNDS!

Figure 7.2   NASA engineer John Houbolt demonstrated that if the Apollo astronauts would perform their rendezvous maneuvers in orbit around the Moon rather than in Earth orbit, the mission would be cheaper, safer, and faster. In his scheme, a small lifeboat craft could be flown down to the surface of the Moon, while its much heavier mother ship would circle overhead. Houbolt's computer simulations showed that, with lunar-orbit rendezvous, the booster rocket that would carry the Apollo astronauts on their deep-space journey would need to weigh only six million pounds—half the weight required to execute the mission using either of the two proposed von Braun schemes.

orbits, we could expect—due to statistical variations—to have a small amount of residual fuel left over on 500 of those flights and a small amount of oxidizer left over on the other 500. Neither fuel nor oxidizer can be burned all by itself because burning requires a mixture of the two.

In order to minimize the average weight of the fuel and oxidizer residuals on the upper stages of the Saturn V, we introduced so-called Propellant Utilization Systems. A Propellant Utilization System employs cleverly designed hardware to monitor the quantities of fuel and oxidizer remaining throughout the flight. It then makes automatic real-time adjustments in the burning-mixture ratio to achieve nearly simultaneous depletion.

For the Saturn V, the necessary measurements were made with capacitance probes running along the length of the fuel tank and the oxidizer tank. A capacitance probe is a slender rod encased within a hollow cylinder. Openings at the bottom of the hollow cylinder allow the fluid level inside to duplicate its level on the outside.

As the fluid level inside the cylinder decreases, the capacitance of the circuit changes proportionately in instant response, thus providing the system with a direct measurement on the amount of fluid remaining in the tank. This continuous fluid-level information is then used in making subtle real-time adjustments in the rocket's burning-mixture ratio to achieve nearly simultaneous depletion of the two propulsive fluids.

THE PROGRAMMED MIXTURE RATIO SCHEME

The Propellant Utilization System on the S-II state increased the performance of the Saturn V enough to allow it to carry 1400 extra pounds of payload to the vicinity of the Moon. But, soon after the first Propellant Utilization System became operational, it created an annoying problem for the mission planning engineers. When we were simulating the cis-lunar trajectories and the corresponding payload capabilities for the Saturn V, we found that if we ran two successive simulations with identical inputs, each simulation would yield a slightly different payload at burnout.

These unexpected payload variations came about because the program's subroutines simulated slightly different statistical variations in the Propellant Utilization System on each flight. In order to circumvent this difficulty, we did what engineers always do—we called a meeting at which we brainstormed various techniques for making the statistical payload variations go away.

Fortunately, no one in attendance that day was able to come up with a workable solution. Sitting in the back of the room was long, lanky propulsion specialist named Bud Brux, who said almost nothing during the meeting. But, when he went back to his office that afternoon, he began to think about the problem we had encountered. Suddenly, Bud Brux said to himself: "Hey, wait a minute! The reason we build a rocket is to put payload into space. If something is causing that payload to vary, perhaps we should accentuate the effect, rather than trying to make it go away."

Bud Brux then wrote a simple, two-page interoffice memo to the trajectory specialists suggesting that we vary the mixture ratio as much as we possibly could to see if this must produce important performance gains. We were not particularly excited by the memo Bud Brux wrote; we received lots of memos in those days. But, when the first few simulations came back from the computer, our excitement shot up by a decibel or two. On the best of those simulations, the Saturn V was able to carry nearly 3000 extra pounds of extra payload to the Moon, each pound of which was worth about $2000—or five times its weight in 24-karat gold!

Figure 7.3 highlights the salient features of the Programmed Mixture Ratio Scheme as applied to the second stage of the Saturn V Moon rocket. Early in that rocket's flight, the burning-mixture ratio was 5.5 to 1 (5.5 pounds of oxidizer for every

THE PROGRAMMED MIXTURE RATIO SCHEME

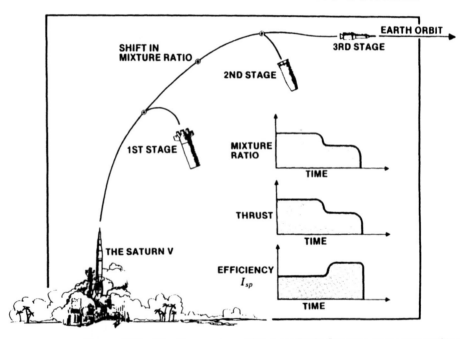

Figure 7.3 The Programmed Mixture Ratio Scheme as suggested by Bud Brux allowed the Saturn V Moon rocket to carry nearly 3000 pounds of extra payload on its translunar journey. This was accomplished quite simply: by adjusting the positions of five valves in midflight. No hardware modifications were required to achieve this important performance gain; the valves had already been installed for other purposes on the rocket stage.

pound of fuel). But 70 percent of the way through the burn, the mixture ratio was suddenly shifted to 4.5 to 1.

As the graphs in Figure 7.3 indicate, this shift in the burning-mixture ratio provides the rocket with high thrust early in its flight, at a slightly lower specific impulse. Then, after the Programmed Mixture Ratio shift, it has lower thrust, but higher specific impulse.

After studying the rocket's behavior, we concluded that the Programmed Mixture Ratio shift causes the rocket to leave more of its exhaust molecules lower and slower as it flies toward the Moon. This, in turn, puts more useful energy into its payload. Another way to gain insight into the rocket's enhanced performance is to study the various trajectory losses associated with a typical Apollo mission.

The ideal velocity for each of the three Saturn V stages is listed in the left-hand column of Table 7.1. These values are followed by the gravity loss, drag loss, and steering loss—each of which is discussed in more general terms in Chapter 4.

The gravity losses for the first two stages are 4000 feet per second and 1100 feet per second, respectively. The third-stage gravity loss amounts to only 400 feet per second. The gravity loss is so low for the third stage because it travels in essentially the horizontal direction as it approaches its destination orbit.

The drag loss for the first stage amounts to 150 feet per second. The two upper stages have no drag losses because, during their burns, the vehicle is traveling through extremely tenuous atmosphere. The first stage flies a "gravity turn" trajectory (thrust vector parallel to its instantaneous velocity vector), so its steering losses are zero. The steering losses for the other two stages are 600 and 15 feet per second, respectively.

The Programmed Mixture Ratio shift provides high thrust (but low specific impulse) in the early part of the flight. This, in turn, allows the optimal steering laws to tilt the rocket into the

TABLE 7.1   Ideal Velocity and Velocity Losses for a Typical Translunar Trajectory

Stage	Ideal Velocity (ft/sec)	Gravity Loss (ft/sec)	Drag Loss (ft/sec)	Steering Loss (ft/sec)
First stage	12,000	4000	150	0
Second stage	15,500	1100	0	600
Third stage	13,500	400	0	15

more horizontal orientation earlier, thus creating important reductions in its overall gravity losses.

Many years ago, I wrote a brief account detailing the important engineering accomplishments of Bud Brux. "If Bud Brux had sent us a map showing us where five solid gold Cadillacs were buried in the company parking lot," I wrote, "that map would not have been worth as much as the memo he actually wrote."

OPTIMAL FUEL BIASING

If we load 1000 identical hydrogen-fuel rockets with the desired amounts of propellants in the proper ratio and then fly all 1000 of them into Earth orbit, approximately 500 will end up with fuel residuals at burnout, and the other 500 will end up with oxidizer residuals.

On the average, of course, the oxidizer residuals will turn out to be approximately five times heavier than the fuel residuals because the rocket carries five pounds of oxidizer for every pound of fuel. Consequently, if we would add a little extra fuel to each of those 1000 rockets before liftoff, that extra fuel would reduce the statistical frequency of the heavier oxidizer residuals. Moreover, when those few oxidizer residuals do occur, they will be lighter because of the fuel biasing.

Figuring out precisely how much extra fuel to add to achieve optimal mission performance turns out to be a difficult problem in statistics. One rather clumsy approach for determining the optimal fuel bias is flowcharted in Figure 7.4. In such a simulation, we command the computer to choose a fuel bias and then sample a value from each of the statistical distribution functions having to do with the variation of the rocket's thrust, its flow rate, its specific impulse, its mixture ratio, and so on.

The computer then substitutes each of these statistical values into our trajectory simulation program, and at burnout, it records the type of residual (fuel or oxidizer) and its corresponding weight. This crude Monte Carlo technique is repeated hundreds or thousands of times so the computer can construct an accurate statistical histogram similar to the one at the bottom of Figure 7.4. Repetitions of the procedure with different fuel-bias levels allow the mission planners to determine which fuel-bias level provides optimum rocket performance.

OPTIMAL FUEL BIASING BY MONTE CARLO SAMPLING

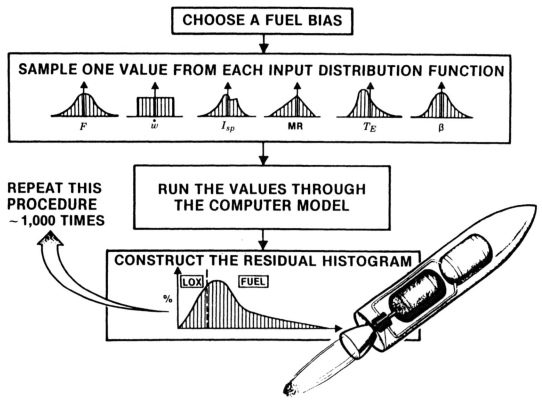

Figure 7.4 Monte Carlo sampling techniques were used in this rather inefficient computational algorithm to determine the optimal fuel bias to add to the Saturn S-II stage, which was designed to burn a 5-to-1 mixture of liquid oxygen and liquid hydrogen. The Monte Carlo algorithms for fuel biasing were rather easy to formulate, but they required approximately 10,000 computer simulations per mission at a cost of about $95,000. Closed-form, fuel-biasing techniques were later perfected. The closed-form approach still required 13 computer simulations per study, but the simulation cost per mission was reduced by approximately 97 percent.

This technique worked as advertised, but it turned out to be extremely expensive, especially in the days when computer simulation time was at a premium. Eventually, however, we worked out a way to achieve optimal fuel biasing using Leibniz's rule for the differentiation of integrals, simple closed-form equations from statistics, and well-known rocketry equations.

For the S-II stage, the Monte Carlo approach typically required 10,000 computer simulations executed at a total cost of $95,000 per flight. The closed-form solution required only 13 computer simulations at a cost of about $3000.

The curves in the lower portion of Figure 7.5, which were constructed using the closed-form equations we derived, can be used to determine the optimum fuel-bias level. For a typical Moon mission, we determined that the optimum amount of fuel to add was about 600 pounds, assuming that we wanted the smallest residual propellant remaining at the 3σ probability level.

Several years later, John Wolfe, a researcher at Rockwell International, figured out how to add the proper amount of fuel bias to *maximize payload* rather than *minimize residuals*. These two bias levels are slightly different because when we add fuel bias to minimize residuals, the fuel bias itself represents a dead weight that the rocket must carry into space. Consequently, biasing to minimize residual propellants is not equivalent to biasing to maximize payload. As the graph in the lower right-hand corner of Figure 7.5 indicates, biasing to maximize payload requires a fuel bias of only about 520 pounds compared with 600 pounds when we added fuel bias to minimize residuals.

OPTIMAL TRAJECTORY SHAPING

The mathematicians in Isaac Newton's day developed a charming custom that helped maintain fruitful interactions with distant colleagues. Whenever one of them managed to solve a particularly difficult mathematical problem, he would delay publication until he had posed it as a challenge to the other expert mathematicians of the day. Such a challenge was issued by the Bernoulli brothers shortly after they managed to solve a rather elegant optimization problem. In their challenge, they gave their fellow mathematicians six months to devise a suitable solution.

One afternoon Isaac Newton heard about the challenge. That night, before he went to bed, he solved the problem, and the next day, he submitted the solution for publication anonymously. Later, when John Bernoulli reviewed the solution, he remarked: "I recognize the lion by his paw." No other mathematician alive at the time, in John Bernoulli's view, could possibly have put together such a clever solution.

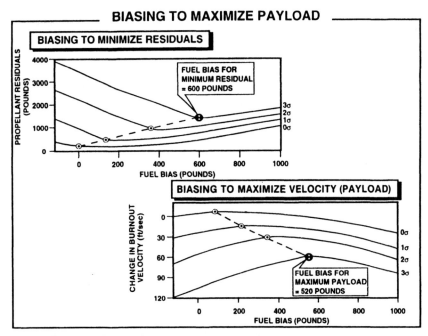

Figure 7.5 As the parametric graphs indicate, the optimal amount of fuel bias needed to minimize the residual weight for the Saturn S-II stage at the 3σ probability level turned out to be approximately 600 pounds. Biasing to maximize payload, which was perfected by Rockwell engineer John Wolfe, required a slightly smaller fuel bias of only 520 pounds.

The problem Newton solved so effortlessly is called the Brachistochrone ("minimum-time") problem. Here is a modified formulation of that problem:

Suppose two fixed points A and B are in a constant one-*g* gravitational field with point A higher than point B. Now suppose we connect point A to point B with a frictionless wire of arbitrary shape. What shape should the wire be so that a frictionless bead will slide from point A to point B in minimum time?

In my first book, *The Rush Toward The Stars*, I could not resist reformulating the Brachistochrone problem in a more picturesque manner by substituting a playground sliding board for the frictionless wire. Notice that Figure 7.6 features five different proposed shapes for the playground sliding board.

The first proposed shape is a straight line that will, of course, provide the minimum possible travel distance for a youngster

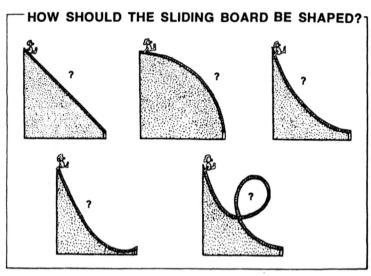

Figure 7.6 The Bernoulli brothers boldly challenged seventeenth-century mathematicians to find the optimal shape for a frictionless playground sliding board that would allow a youngster to slide from top to bottom in minimum time. Isaac Newton solved the problem in a single afternoon. Later, when he published the solution, John Bernoulli was so impressed with its cleverness, he is said to have observed: "I recognize the lion by his paw."

sliding down its slope. But will this simple shape also provide the minimum travel time? The second proposed shape involves a convex contour bowed upward. The third and fourth are concave downward. Finally, a fifth proposed shape includes a hazardous loop so the youngster will flip over and fall on his head during his adventuresome ride.

Isaac Newton proved mathematically that the optimal shape for the sliding board under the stated conditions is a specific contour called a "cycloid." As the overlapping sketches in Figure 7.7 indicate, a fixed point on the rim of a wheel that is rolling on a flat surface traces out the cycloid in question. Newton's cycloid is vaguely reminiscent of the celestial epicycles Ptolemy relied on 2000 years ago when he attempted to explain the curious motion of the wandering stars (see Chapter 1).

Figure 7.8 highlights the travel times associated with three sliding boards with distinctly different shapes. In each case, the sliding board will transport the youngster from coordinate (0, 0) to coordinate (100, 100) in a different amount of time.

Rotating Wheel Creating the Brachistochrone Cycloid

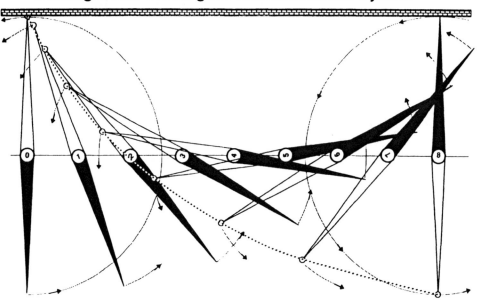

Figure 7.7 A point marked on the rim of a wheel rolling along a flat surface generates this special curve called a cycloid. That same cycloid shape turns out to be the optimum contour for a frictionless playground sliding board that will allow an enthusiastic youngster to slide from top to bottom in minimum time.

The total sliding time for the frictionless, straight-line sliding board is 3.53 seconds and the sliding time for the one contoured along a circular arc is 3.27 seconds. The optimal cycloid, which roughly splits the difference between the other two shapes, provides the minimum sliding time—3.21 seconds.

Notice that the dashed-line curve running across the top of Figure 7.8 is labeled "the leaping lizard." The leaping lizard is a creative youngster who provides us with an interesting new way of attacking optimization problems. Rather than choosing to slide down one of the available sliding boards hoping to win the race, he gets a big run to hurl his body horizontally into the air with just enough velocity so he will crash into the bottom of the sliding board.

As the simple calculations on the right-hand side of Figure 7.8 indicates, the leaping lizard travels from point A to point B in only 2.5 seconds! Of course, he is breaking one of the rules of

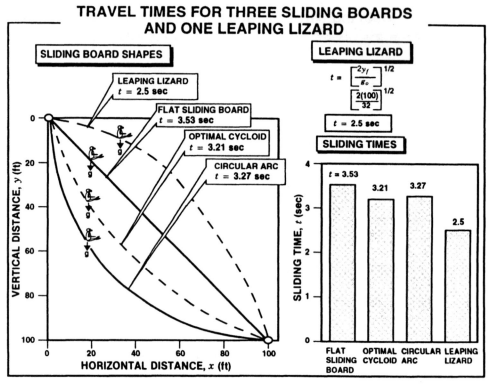

Figure 7.8 A frictionless, cycloid-shaped sliding board allows a youngster to slide down 100 feet with a horizontal travel distance of 100 feet in the shortest time (3.21 seconds) of any shape. For comparison purposes, the figure also presents the corresponding sliding times associated with a flat sliding board and a sliding board curved along a circular arc. The "leaping lizard" is a clever youngster who abandons the use of sliding boards entirely. He leaps sideways with just enough velocity so his body will crash into the bottom of the sliding board. Using this imaginative technique, the "leaping lizard" can reach the bottom in only 2.5 seconds.

the original problem formulation: He is not using the steady force of gravity to move his body along the curved surface of a sliding board.

However, the leaping lizard is teaching us an important lesson about optimization problems: If we are allowed to break some of the constraints in a problem formulation, we can sometimes discover a substantially more efficient way to carry out an important operation or process. If you are assigned to devise some specific optimal solution, you should consider emulating the courage and creativity of the leaping lizard. Sometimes you can beat everyone else if you bend or break some real or apparent constraint.

This courageous approach can yield important performance gains in connection with low-thrust spiraling trajectories. Specifically, if an astronaut is in a 100-nautical-mile parking orbit and she wants to reach escape velocity at a constant acceleration of 0.001 g, she will have to spiral around Earth until her space ship builds up an effective velocity increment of about 19,000 feet per second.

However, if she can break just one of the apparent constraints and turn her rocket off and on repeatedly, she can drive her spacecraft onto an escape parabola with only about 14,000 feet per second equivalent ΔV. If you are assigned to handle some specific optimization problem, you should always explore various possibilities for breaking one or more of the constraints to see if substantial performance gains can be achieved.

The sketch in Figure 7.9 represents a typical trajectory for the space shuttle when it is boosting itself from the surface of Earth to an altitude of 90 n.mi. Notice how the trajectory shape is vaguely similar to the solution of the Brachistochrone problem Isaac Newton so skillfully derived nearly 300 years ago. Indeed, some optimal rocket trajectories, in essence, are inverted Brachistochrone (minimum time) contours in which gravity must be overcome instead of providing a motive force. All other things being equal, if a rocket's ascent trajectory can be shaped so its payload reaches its destination orbit in a shorter time, that rocket can carry a heavier payload into space. Moreover, the mathematical algorithms used in solving both the Brachistochrone and trajectory optimization problems are essentially the same.

POSTFLIGHT TRAJECTORY RECONSTRUCTION

On January 1, 1801, the first minor planet, Ceres, was spotted by alert astronomers as it hooked around the Sun. Ceres, which we now call an asteroid, was a new type of object never seen by anyone up until that time. Unfortunately, after Ceres had been in view for only 41 days, it traveled so near to the Sun it was lost from view. The astronomers who were tracking it were afraid that it might never be found again.

However, as Figure 7.10 indicates, the great German mathematician, Karl Frederich Gauss, accepted the challenge of trying to reconstruct the trajectory of Ceres from the small number of astronomical observations available to him. Under his brilliant direction, Ceres was located again on the other side

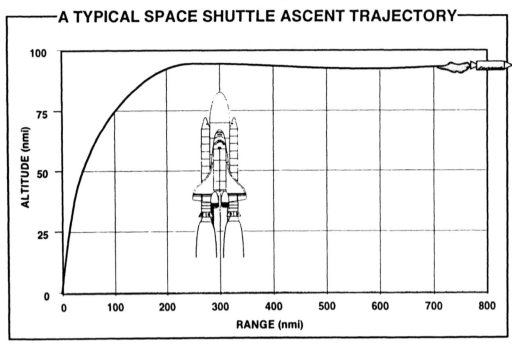

Figure 7.9 When a rocket rises from the surface of Earth bound for a low-altitude orbit, its trajectory can be regarded as an inverted Brachistrochrone (minimum-time) path. This is true because if we can minimize the travel time for the rocket—all other things being equal—it will consume less propellant during its journey into space.

of the Sun on the last day of 1801, almost exactly one year after it had first been discovered.

The mathematical methods Gauss used in reconstructing the orbit of Ceres are still used today for postflight trajectory simulations to determine the actual performance of large liquid-fueled and solid-fueled rockets. When we are executing a *preflight* trajectory simulation, we feed the thrust and flow-rate profiles into the program together with the initial weight of the vehicle, its guidance angle histories, and the like, and then we simulate the resulting trajectory of the rocket. In a *postflight* trajectory simulation, we do just the opposite. We feed the program with the trajectory of the rocket—as ascertained by the tracking and telemetry measurements—and then we use the computer to determine the thrust and flow-rate

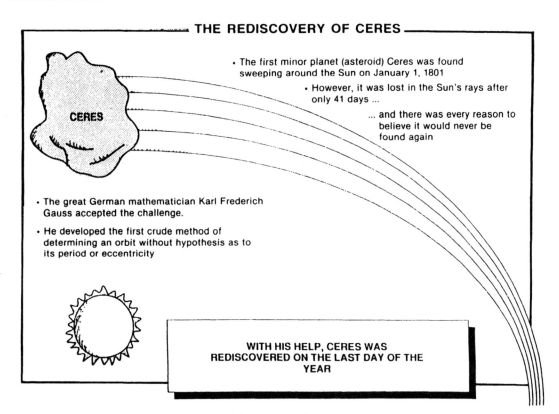

THE REDISCOVERY OF CERES

CERES

- The first minor planet (asteroid) Ceres was found sweeping around the Sun on January 1, 1801
- However, it was lost in the Sun's rays after only 41 days ...
- ... and there was every reason to believe it would never be found again
- The great German mathematician Karl Frederich Gauss accepted the challenge.
- He developed the first crude method of determining an orbit without hypothesis as to its period or eccentricity

WITH HIS HELP, CERES WAS REDISCOVERED ON THE LAST DAY OF THE YEAR

Figure 7.10 In 1801, the German mathematician Karl Frederic Gauss developed a new mathematical algorithm called "the iterative least squares procedure" to reconstruct in exquisite detail the heliocentric trajectory of the asteroid Ceres. Ceres had vanished behind the Sun earlier that year. But, by years end, contemporary astronomers pointed their telescopes toward a tiny patch of sky—under the direction of Gauss—to rediscover Ceres as it hooked around the back side of the Sun.

profiles and the guidance angles the booster must have had in order to have traveled along the observed trajectory.[1]

Those of us who worked as trajectory experts on the Saturn V Moon rocket developed one of the most complicated and sophisticated postflight trajectory reconstruction programs

[1] In a television interview on ABC TV, I was once asked what a trajectory expert does for a living. "We predict where the rocket will go before the flight," I replied. "Then, after the flight, we try to explain why it didn't go there."

ever formulated until that time. It included more than 10,000 lines of computer code and it required 600 discreet inputs per simulation, all of which had to be accurate and correct if the program was to produce the desired results. Unfortunately, approximately 75 percent of our simulations blew up due to erroneous inputs.

In a typical postflight reconstruction, we simulated a 400-second trajectory segment. This required about 2.5 hours of computer time on an IBM 7094 mainframe computer at a cost of about $700 per hour. The six-degree-of-freedom iterative least squares hunting procedure was structured so we could, on any given simulation, choose up to seven independent variables, such as attitude, slant range, inertial velocity, and the like. We could choose up to seven dependent variables, such as the thrust profile, the flow-rate history, the initial weight of the rocket, and so on.

The dependent variables were measured during the flight with tracking instruments on the ground and telemetry devices carried onboard the rocket. On a typical Saturn V trajectory reconstruction, the computer calculated about 30 partial derivatives at each of 400 time points spaced one second apart. The resulting partial derivatives—around 12,000 of them— were arranged in matrix format and recorded on seven magnetic tapes.

On a typical flight, the average deviation between the predicted preflight trajectory and the actual postflight trajectory was approximately one mile. However, after 2.5 hours of simulation time on the computer, the iterative least squares hunting procedure typically reduced this average error to only about one foot.

After running a series of computer simulations of this type, we were able to reduce the statistical tolerances on the rocket's performance parameters such as thrust and specific impulse. This, in turn, allowed us to improve the performance capabilities of the rocket by several hundred pounds of payload headed for the Moon.

SUMMARY OF PERFORMANCE GAINS

Table 7.2 summarizes the magnitudes of some of the performance enhancements we were able to achieve for the Saturn V Moon rocket. In optimal fuel biasing, we added a little extra fuel to reduce the statistical frequency of the heavier

TABLE 7.2 Schemes for Enhancing the Performance Capabilities of the Saturn V Moon Rocket

Approach	What Produces the Performance Gain?	Performance Gain (pounds of extra translunar payload)	Comments
Optimal Fuel Biasing	By adding a small amount of extra fuel for the rocket, we can reduce the statistical frequency and the average mass of the heavier oxidizer residuals	400	The optimal fuel bias turns out to be about 600 pounds of extra fuel
The Programmed Mixture Ratio Scheme F [graph] TIME	By introducing a sudden shift in the burning-mixture ratio in midflight, we can leave more of the exhaust molecules lower and slower, thus increasing the vehicle's payload capabilities	2700	The optimal location for the mixture ratio shift occurs when approximately 70% of the propellants have been burned
Postflight Trajectory Reconstruction	The iterative least squares trajectory reconstruction program automatically determines the thrust, flow rate, etc., on that particular flight. Over a series of flights, this information can be used to reduce the flight performance reserves	800	The 6-degree-of-freedom trajectory reconstruction program includes 10,000 lines of computer code
	Total	3900	The translunar payload equals approximately 100,000 pounds. The resulting 3.9% payload gain was worth about $84,000,000 per flight. This performance gain was achieved without changing any hardware

oxidizer residuals. This allowed the rocket to carry 400 extra pounds of payload to the Moon. The resulting gain was achieved by adding approximately 600 extra pounds of fuel bias on each lunar flight.

The Programmed Mixture Ratio Scheme, which involved a sudden shift of the burning-mixture ratio, allowed the rocket to leave its exhaust molecules lower and slower, thus putting more energy into the payload. This technique, required no hardware modifications, yet it allowed the vehicle to carry 2700 extra pounds of translunar payload.

Postflight trajectory reconstruction, which relied on the iterative least squares hunting procedure, allowed us to reduce the statistical tolerances on various important measures of rocket performance, such as thrust and flow-rate profiles and guidance angles. In accomplishing this, we employed a six-degree-of-freedom trajectory simulation program coupled with an iterative least squares hunting procedure that included 10,000 lines of computer code. Postflight trajectory reconstruction allowed the rocket to carry 800 extra pounds of payload on its cis-lunar journey.

Taken together, these three performance-enhancement techniques produced payload gains totaling 3900 pounds, each pound of which was worth about five times its weight in 24-carat gold. These gains were achieved on all of the Apollo Moon flights without costly hardware modification through creative engineering and careful analysis techniques. Taken together, those hard-won payload gains provided most of the extra performance margins that were needed to allow Neil Armstrong and Edwin Aldrin to frolic across the lunar surface.

8

CHOOSING THE PROPER ORBIT FOR A SATELLITE

Newton dived into Nature's hidden springs and laid bare the
principles of things and gave us worlds unknown before.
—*Adapted from Charles Churchill's* Webster's New World
Dictionary of Quotable Quotes

In the earliest days of the U.S. space program, our country's
aerospace experts were delighted if they were able to launch a
satellite into any orbit that did not intersect Earth. But over the
years, aerospace technology has improved dramatically so that,
today, we often target our satellites for extremely precise orbital locations, some of which have special geometrical properties or useful environmental characteristics. These special
destinations in space include polar orbits, Sun-synchronous
orbits, geosynchronous and Molniya orbits, and stable libration-
point configurations—all of which are discussed in the sections to follow.

POLAR ORBITS AND POLAR BIRDCAGE CONSTELLATIONS

In 1957, when the first Russian Sputnik began spiraling around
Earth, a group of research scientists at the Applied Physics
Laboratory at Johns Hopkins University found a clever way to
determine Sputnik's orbital elements using measurements of
the Doppler-shift variations associated with its broadcast
signals.

Shortly thereafter, one of them, Frank McClure, realized that
this technique was reversible. It should be possible, he con-

cluded, to use similar real-time, Doppler-shift measurements to determine the positions of users situated on the ground. This simple observation eventually came to fruition in the form of the six-satellite Transit navigation constellation, which is shown in Figure 8.1. Notice that all six of the Transit satellites are launched into 580-nautical mile polar "birdcage" orbits that carry each satellite alternately over the North Pole and the South Pole.

The Transit Navigation Satellites

The computational algorithm used in determining the location of a receiver that is picking up the signal from a Transit navigation satellite is conceptually akin to the technique a clever navigator might use to establish his location relative to a pair of railroad tracks. Imagine a train moving along the tracks, blowing its whistle as it goes. A navigator standing beside the tracks, who measures the frequency of the sound waves coming from

THE TRANSIT SATELLITES ARE LAUNCHED INTO POLAR "BIRDCAGE" ORBITS

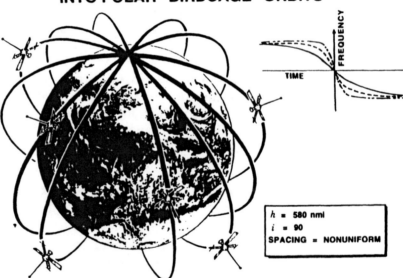

h = 580 nmi
i = 90
SPACING = NONUNIFORM

Figure 8.1 All of the Transit navigation satellites have been launched into a constellation of so-called polar birdcage orbits 580 nautical miles above Earth. Special receivers on the ground measure the real-time, Doppler-shift variations from one of the transit satellites as it travels across the sky. This information is then used to figure out the user's longitude and latitude in a real-time navigation solution.

the moving train, would observe an increase in the frequency as the train approaches and a corresponding decrease as it passes by and recedes into the distance. His graph of observed frequency versus time would be a "step function": high frequency at first, followed immediately by low frequency.

If he then moves three miles away from the tracks and constructs a similar graph of frequency versus time, it will not be a step function. It will be, instead, a gentle S-shape curve. The farther he moves away from the tracks, the more gentle will be the S-shape curve.

This simple, conceptual technique forms the basis of the solution algorithm employed by the users of the Transit navigation system. When a Transit satellite sweeps from horizon to horizon, real-time measurements of its Doppler shift curve provide direct information on the user's distance from its current ground track. In addition, timing information and satellite ephemeris constants modulated onto the sinusoidal carrier waves coming down from the satellite allow the user to determine where he is along the ground track. These two pieces of information are then computer-processed to ascertain the user's current longitude and latitude. Polar birdcage orbits were selected for the low-altitude Transit satellites to provide rapidly varying Doppler-shift measurements, global (but intermittent) coverage, and to achieve stable inertial nodes (zero nodal regression rates).

The designers of the Transit navigation system realized that practical navigation services would be possible only if each Transit satellite could be oriented so its navigation antennas would always point toward the center of Earth. This could, of course, be accomplished by firing attitude control thrustors repeatedly to maintain a constant Earth-seeking orientation. However, this approach would consume valuable propellants, thus shortening the life of the satellite.

The Transit design team managed to perfect a clever alternative for maintaining an Earth-seeking orientation—without using propellants. It is called *gravity-gradient stabilization*. First, they launch a Transit satellite into orbit and orient it so its navigation antenna points vertically downward toward the center of Earth. Then they erect a 50-foot telescoping boom vertically upward away from the center of Earth. The gravitational force acting on the lower edge of the spacecraft is then stronger than the force on its upper edge. This so-called gravity gradient induces restoring torques that help maintain the satellite in a vertical orientation as it travels around Earth.

The Iridium Constellation

The Iridium constellation of mobile communication satellites being masterminded by Motorola engineers will consist of 66 satellites launched into polar birdcage orbits. Those 66 satellites will travel over the North Pole and South Pole at an altitude of 413 nautical miles in six orbit planes, with 11 satellites per plane.

The Iridium satellites, working together in partnership, will provide supplemental cellular telephone coverage for users all around the globe. Motorola is slated to spend $3.6 billion to build and launch this rather ambitious constellation of satellites.

SUN-SYNCHRONOUS ORBITS

A different group of near-polar orbits, called Sun-synchronous orbits, have another highly desirable geometrical property: Their nodal regression rates match the rate at which Earth travels around the Sun (approximately 1° per day). As Figure 8.2 indicates, the orbit of a satellite coasting around a perfectly spherical Earth will maintain an orientation that is fixed in inertial space. This is indicated in the left-hand side of the figure in which all the orbit planes are essentially parallel to one another.

The angle between the orbit plane and the radius vector to the Sun is called the β (beta) angle. For most orbits, the β angle varies continuously throughout the year. However, we can take advantage of the systematic perturbations induced by Earth's equatorial bulge to create a satellite orbit that twists approximately 1° per day, thus canceling most of the angular variation that would otherwise occur as Earth travels around the Sun. If the edge of the orbit plane points toward the Sun initially, it will continue to point toward the Sun (to a first approximation) throughout the life of the satellite. This is indicated on the right-hand side of Figure 8.2.

A Sun-synchronous satellite does not have to be launched into an orbit with the edge of the orbit plane pointed toward the Sun (β angle = 0°); any β angle can be maintained (approximately) in this manner. A so-called dawn–dusk orbit (β angle = 90°) is shown in the upper left-hand corner of Figure 8.3. In this case, the orbit plane remains essentially *perpendicular* to the Sun. Throughout the year, of course, the Sun moves from the Northern Hemisphere to the Southern Hemisphere, swing-

ORBITAL MECHANICS FOR SUN-SYNCHRONOUS ORBITS

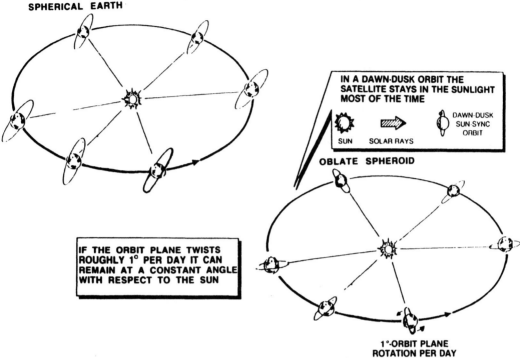

Figure 8.2 A satellite traveling around a spherical Earth with no other forces of perturbation will trace out a series of orbits all of whose planes are essentially parallel to one another in inertial space. But if we launch that same satellite into a Sun-synchronous orbit around an oblate spheroid with a carefully chosen altitude-inclination combination, its orbit plane can be made to twist at a rate of 360° per mean solar year. When combined with repeating ground-trace geometry, this approach provides nearly constant solar illumination of Earth's surface to make the satellite's observations easier to interpret.

ing above and below the equator approximately 23.5°. This north–south movement induces corresponding variations in the β angle throughout the year.

Sun-synchronous orbits provide essentially constant lighting conditions for weather satellites and Earth resources satellites. Such satellites must be launched into orbits with particular altitude-inclination combinations; otherwise, the orbit plane would not twist at the proper rate. The graph in Figure 8.3 represents the locus of all circular Sun-synchronous orbits lying below an altitude of 1000 nautical miles. All Sun-synchronous orbits have orbital inclinations greater than 90° so they will twist in the proper direction to follow the Sun.

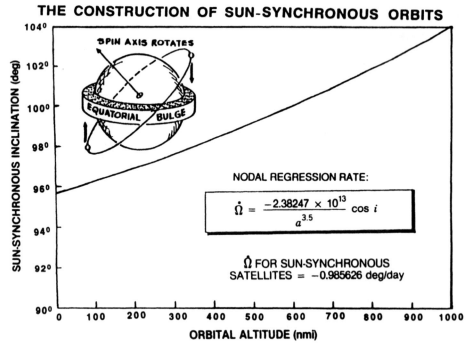

Figure 8.3 The gently curving line running across this graph represents the locus of all circular Sun-synchronous orbits lying below a 1000-nautical-mile altitude. Those in low-altitude orbits (at 100 nautical miles or so) must have orbital inclinations of approximately 96°. Those in higher-altitude orbits must have higher inclinations because Earth's equatorial bulge has an increasingly smaller effect on satellites in the high-altitude flight regime.

A Sun-synchronous satellite in a 100-nautical-mile circular orbit must have an orbital inclination of about 96°. At higher altitudes, Earth behaves more like a point mass, so the circular Sun-synchronous orbits at these higher altitudes must have even higher inclinations. At 1000 nautical miles, for instance, the inclination of a circular Sun-synchronous orbit must be approximately 104°. The Landsat D Earth resources satellite was positioned in a 380-nautical-mile orbit, at an inclination of 98.2°. Its orbital geometry was purposely chosen so its equatorial crossings (descending nodes) would all occur at 9:45 A.M. local time.

The nodal crossing point for the Landsat D was purposely selected to provide slanting rays of the Sun for its observations, which cause the surface features to stand out in high relief. Researchers have found that shadows of intermediate length best highlight the terrain contours for accurate photo interpretation.

FULL-SUN, SUN-SYNCHRONOUS ORBITS

A near-polar satellite launched into a Sun-synchronous orbit with an angle of approximately 90° is said to occupy a dawn-dusk orbit. To a first approximation, it travels continuously along the day–night terminator line. A satellite in any dawn-dusk, Sun-synchronous orbit would be in sunlight continuously if the Sun did not move above and below the equator as the seasons change.

Under the proper conditions, a satellite traveling around a dawn–dusk orbit can remain in sunlight 100 percent of the time (see Figure 8.4). Notice that Figure 8.4 includes a mid-winter ray from the Sun tangent to the surface of Earth. The curving arc being pierced by that solar ray represents the locus of all dawn–dusk, Sun-synchronous orbits in polar coordinates (altitude versus inclination). The dawn–dusk, Sun-synchronous satellites that will remain in sunlight 100 percent of the time all occupy an altitude range that lies between 750 and 1800 nautical miles.

Notice that the locus line defining the Sun-synchronous orbits gradually bends toward the equator. This relentless trend gives rise to an obvious question: Is there some upper limit on the altitude of a circular Sun-synchronous orbit? Such an upper limit indeed does exist; it equals 3226 nautical miles. At any altitude higher than 3226 nautical miles, circular Sun-synchronous orbits around Earth do not exist.

GEOSYNCHRONOUS ORBITS

Most orbiting satellites whiz around Earth at a high rate of speed. A satellite launched into a low-altitude "treetop" orbit, for example, will go all the way around the world in less than 90 minutes. By contrast, the Moon, which is also an Earth-orbiting satellite, lumbers around Earth in 28.5 days.

Intermediate between these two altitude extremes is a special orbit 19,300 nautical miles high. A satellite launched into a circular equatorial orbit at that particular altitude will hang motionless in the sky as seen from the spinning Earth.

Inclined Geosynchronous Orbits

A satellite launched into an *inclined* circular orbit at the 19,300-nautical-mile altitude will trace out a figure-8 ground trace every 24 hours. If we launch it into an orbit with a 60°

SUN-SYNCHRONOUS ORBITS WITH CONSTANT SOLAR ILLUMINATION (DAWN–DUSK ORBITS ONLY)

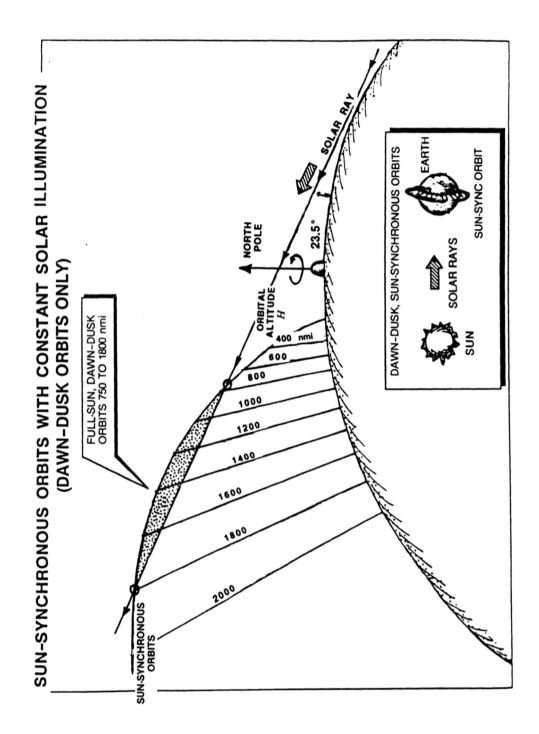

inclination (see Figure 8.5) its north–south latitude extremes will equal 60°. The sine-curve-shape contour on the Mercator projection in Figure 8.5 corresponds to the ground track that would be traced out by a satellite in an inclined 19,300-nautical-mile orbit over a *nonrotating* Earth.

In the real world, of course, Earth rotates out from under the satellite's orbit plane at 15° per hour. Each one of the black or gray stripes in Figure 8.5 represents a two-hour interval during which Earth rotates 30° in the eastward direction. Notice how Earth's rotational motion, combined with the motion of the satellite, produces the figure-8 ground trace in Figure 8.5.

An early communication satellite called Syncom II was purposely launched into a 19,300-nautical-mile orbit with a 28.5° orbital inclination. This strange choice of orbits had to be made because the booster rocket did not have enough performance capability to make the necessary plane change. Syncom II's successor, Syncom III, was launched by a rocket with improved performance, and it was lofted into a true geostationary orbit. Dozens of satellites, most of which are used for communication relay, are now orbiting at the geosynchronous altitude. They form a thin, metal daisy chain above the equator.

The Perturbations Acting on a Geosynchronous Satellite

A geosynchronous satellite will remain over a fixed spot on the equator only to a first approximation. Such satellites are pushed away from their stationary locations by three major perturbations:

1. The tesserial harmonic
2. Lunar–solar perturbations
3. Solar-radiation pressure

Figure 8.4 This polar-coordinate graph depicts the locus of all Sun-synchronous orbits and their spatial locations when they are in the vicinity of the North Pole. The straight slanting line represents a single sunbeam passing over the North Pole in midwinter when the Sun is 23.5° south of the equator. The shaded region highlights those specific dawn–dusk Sun-synchronous orbits that are illuminated by the Sun at all times. For continuous solar illumination, such a satellite must be launched into a circular dawn–dusk orbit at altitudes lying somewhere between 750 and 1800 nautical miles.

COMPONENTS OF MOTION THAT MAKE UP THE FIGURE-8 GROUND TRACE FOR AN INCLINED GEOSYNCHRONOUS ORBIT

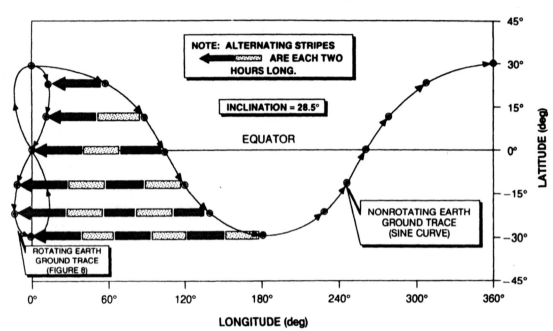

Figure 8.5 This graph shows why a satellite launched into a 24-hour circular orbit 19,300 nautical miles high will trace out the same figure-8 ground trace repeatedly every 24 hours. The sine-curve-shape contour depicts the ground trace of the satellite over a nonrotating Earth. The black and gray stripes each represent a two-hours interval during which Earth rotates out from under the satellite's orbit plane at 15° per hour. As these sketches indicate, the figure-8 ground trace results from the motion of the satellite around its orbit combined with the rotation of Earth.

The *tesserial harmonic* is a higher-order-shape term of Earth. Its effect on geosynchronous satellites arises because if we would slice through Earth at the equator with a big meat cleaver, the resulting cross-section would not have a circular shape. Instead, it would be shaped like an ellipse. As Figure 8.6 indicates, this gravitational anomaly causes a geosynchronous satellite to drift either in the eastward or westward direction, depending on its initial longitudinal location along the geosynchronous arc. The figure highlights four stable longitudinal locations. A satellite positioned at any of these four locations will drift in neither the eastward nor westward direction. For-

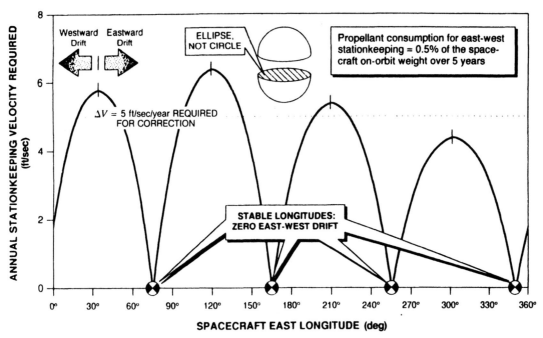

ANNUAL EAST-WEST LONGITUDINAL STATIONKEEPING VELOCITY FOR GEOSYNCHRONOUS EQUATORIAL ORBITS

Figure 8.6 A satellite positioned along the geosynchronous arc usually drifts either eastward or westward due to perturbations induced by the nonsymmetrical gravitational harmonics of Earth. Satellites that are at latitudes on the left-hand side of one of the four bumps in this figure drift in the eastward direction. Those to the right of one of the bumps drift toward the west. The total ΔV required to cancel out these gravitationally induced perturbations amounts to, at most, only about seven feet per second per year. The four points along the bottom of the figure correspond to initial longitudinal locations at which the satellite will drift neither in the eastward direction nor westward.

tunately, the east-west drift of a geosynchronous satellite at any longitudinal location can be controlled at a ΔV cost of less than seven feet per second per year.

The *lunar-solar perturbations*, which are created by the gravitational forces from the Moon and the Sun, induce systematic variation in the inclination of a geosynchronous satellite (see Figure 8.7). If the satellite is launched into an orbit with a 0° inclination, it will move farther and farther above and below the equator along an ever-increasing figure-8 ground trace throughout its useful life. If no correction burns are executed, the satellite's orbital inclination will reach 8° at the end of 10 years on orbit. This linear increase in the inclination angle is quantified by the small graph in the left-hand side of Figure 8.7. Periodic on-orbit corrections for this so-called north–south drift requires a total ΔV of 158 feet per second per year.

Under certain conditions, an orbital inclination of 4° or less can be maintained over a 10-year mission life without using any corrective ΔV's. This can be accomplished by launching the satellite at a particular time of day to achieve the proper orbital geometry with respect to the Sun and the Moon. When this technique is being used, the satellite is purposely launched into an orbit with a 4° orbital inclination. As the graph in the

NORTH–SOUTH STATIONKEEPING ΔV NEEDED TO MAINTAIN AN ACCEPTABLE SMALL INCLINATION FOR A GEOSYNCHRONOUS ORBIT

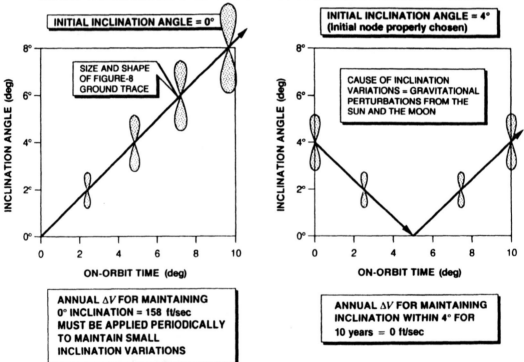

Figure 8.7 The gravitational attractions of the Sun and the Moon induce inclination variations in a geostationary satellite's orbit, thus causing it to move mostly in the north–south direction along a slender figure-8 ground trace. If a satellite starts its mission with a 0° orbital inclination, its inclination will increase over the next several years at an average rate of about 0.8° per year. If desired, onboard rockets can be fired periodically to cancel this so-called north–south drift for a total velocity increment of 158 feet per second per year.

right-hand side of Figure 8.7 indicates, this relatively small orbital inclination will shrink down to 0° for the first five years of the satellite's on-orbit life, then it will increase for the next five years.

The *solar-radiation pressure* acting on a geosynchronous satellite induces systematic variations in its orbital eccentricity. For six months, the eccentricity builds up relentlessly, as shown in Figure 8.8. Then, over the next six months, it shrinks down to zero again. The pressure induced by solar illumination impinging on a surface one astronomical unit

SOLAR-RADIATION PRESSURE PERTURBATIONS ACTING ON A GEOSYNCHRONOUS SATELLITE

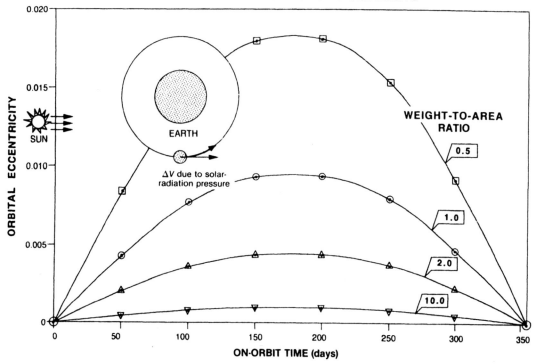

Figure 8.8 The solar-radiation pressure induced by the Sun shining on a satellite at most, amounts to only about five pounds per square mile. However, because this seemingly small force of perturbation is so steady and unremitting, it can create relatively large variations in the eccentricity of a geosynchronous satellite. Such a satellite with a W/A (weight-to-area ratio) of 0.5 pound per square foot, for example, will experience an orbital eccentricity variation peaking at about 0.018. As this graph indicates, the eccentricity of the satellite's orbit increases for six months and then it gradually decreases throughout the next six months.

from the Sun amounts, at most, to only about five pounds per square mile. However, despite its seemingly small magnitude, such a force acting relentlessly over such a long period of time can significantly perturb the orbit of a geosynchronous satellite.

The peak variation in eccentricity induced by solar-radiation pressure is a direct function of the satellite's weight divided by the cross-sectional area it presents to the Sun. As Figure 8.8 indicates, a dense and heavy satellite (bowling ball) will experience relatively small eccentricity variations, whereas one with a lower density (volleyball) will experience considerably larger eccentricity variations throughout the year.

Nullifying the Effects of the Perturbations

International regulatory bodies specify the assigned orbital slot (longitudinal location) of each geosynchronous satellite. Consequently, as Table 8.1 indicates, periodic corrections for east–west drift are almost always executed. Fortunately, the necessary corrective ΔV totals less than seven feet per second per year. Sometimes, in addition, those who own and operate geosynchronous satellites provide the corrective ΔV of 158 feet per second per year needed to execute accurate North-South corrections. Others with less demanding requirements allow their satellites to drift north and south. Ground-based users typically pay lower rates for the use of a satellite that drifts along a north–south figure-8 ground trace compared with one that is maintained at a stationary location over Earth's equator.

Solar-radiation pressure creates orbital eccentricity variations, thus causing the satellite's orbit to have a varying oblate shape. Corrections for orbital eccentricity variations would typically cost several hundred feet per second per year with little practical benefit. Consequently, for most missions, no attempt is made to nullify the eccentricity variations.

12-HOUR SEMISYNCHRONOUS ORBITS

The satellites in the Navstar Global Positioning System (GPS) constellation, those in the Soviet Glonass constellation, and a few other individual satellites are launched into medium-altitude orbits. The Navstar GPS satellites, for instance, occupy 12-hour semisynchronous orbits with 55° inclinations and orbital altitudes of 10,898 nautical miles. That particular

TABLE 8.1 Systematic Perturbations Acting on the Orbit of a Geosynchronous Satellite

Source of the Perturbation	Effect of the Perturbation	Stationkeeping ΔV Necessary to Correct for the Perturbation	Types of Corrections Usually Made
Noncircular cross-section of Earth's equator creates gravitational perturbations.	The satellite appears to drift eastward or westward along the equator.	The required velocity change varies with the satellite's assigned orbital position above the equator. The total ΔV is typically 7 ft/sec/year or less.	East–west position corrections are usually made to avoid having the satellite intrude on the orbital slots of other, nearby satellites.
Gravitational perturbations from the Sun and the Moon distort the satellite's orbit	The inclination of the satellite's orbit changes. This creates an increase or a decrease in the size of its figure-8 ground trace.	Maintaining a 0° inclination requires a ΔV of 158 ft/sec/year.	North–south corrections are usually made unless moderate inclination variations can be tolerated. The Comsat "nodding" maneuver can be used to minimize on-orbit propellant consumption.
Solar radiation pressure from the Sun's rays systematically perturbs the satellite's orbit.	The satellite's orbit experiences systematic changes in eccentricity. The eccentricity builds up for the first six months and then shrinks during the next six months.	The required velocity change for continuously maintaining a circular orbit varies with the satellite's ballistic parameter. Typically, the total ΔV amounts to a few hundred ft/sec/year.	Eccentricity corrections are usually unnecessary because small variations in the shape of the orbit can be tolerated for most missions.

altitude-inclination combination produces a 12-hour orbital period and a one-day repeating ground trace for each GPS satellite.

Figure 8.9 pins down the orbital locations of the 24 semi-synchronous GPS satellites at a particular instant in time. Notice how the various satellites occupy six 55° orbit planes with their nodal crossing points equally spaced around the equator. Each Block II GPS satellite weighs about 2000 pounds. It is constructed at a cost of about $60 million from 65,000 separate parts, yet it is designed to last 7.5 years, or 580 million miles, whichever comes first.

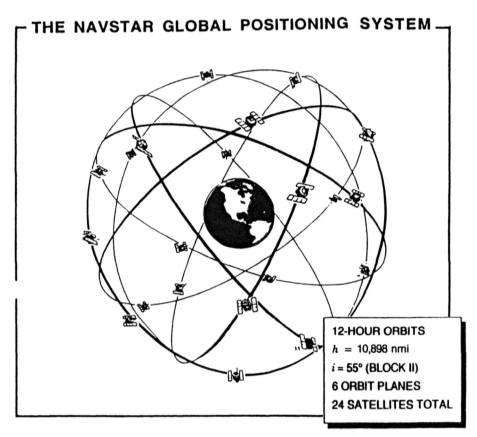

THE NAVSTAR GLOBAL POSITIONING SYSTEM

12-HOUR ORBITS
h = 10,898 nmi
i = 55° (BLOCK II)
6 ORBIT PLANES
24 SATELLITES TOTAL

Figure 8.9 The Navstar GPS satellites are targeted for circular 10,898-nautical-mile orbits with orbital periods of approximately 12 hours. The 24 satellites in the Navstar constellation are positioned in six orbital rings with four satellites in each ring. The inclination of each ring is 55° with the six inertial ascending nodes spaced 60° apart around the equator.

RUSSIA'S MOLNIYA COMMUNICATION SATELLITES

Once they managed to develop large booster rockets with sufficient payload-carrying capabilities, American communication specialists became enamored with geosynchronous satellites because such satellites could be made to hover motionless in the sky. This choice greatly simplified mission planning and satellite tracking.

Soviet rocket scientists usually select a completely different kind of orbit for their communication satellites. Their choice of orbits has been motivated by the need to provide coverage for the many Soviet citizens who live at extremely high northern latitudes. A geostationary satellite seen from an extreme northern latitude tends to lie uncomfortably close to the horizon where blockages and signal distortions routinely occur.

To better serve population centers situated in the far north, Soviet communication satellites are often launched into elongated elliptical *Molniya* orbits with 12-hour periods and 63.4° orbital inclinations. A satellite coasting around such an orbit automatically maintains a zero apsidal rotation rate. Consequently, its apogee can be positioned so it will always remain above its northernmost latitude.

Because they are launched into highly elliptical orbits, Molniya communication satellites travel rather slowly when they are moving across the Northern Hemisphere. In fact, such a satellite spends about 90 percent of its time north of the equator. The ground-trace geometry for a typical Molniya satellite is shown in Figure 8.10. The small circles rimmed around the satellite's ground trace represent the variation in the apparent size of Earth as it would appear when viewed from the satellite. Notice that the ground trace of each Molniya satellite includes a double loop. This complicated ground trace is created by the continuous variations in the position and velocity of the satellite and the steady rotation of the spinning Earth.

ACE-ORBIT CONSTELLATIONS

The left-hand side of Figure 8.11 highlights an important difficulty associated with the Russian Molniya satellites. The Molniya orbits are inclined 63.4°, so the line of apsides does not rotate throughout the year. Consequently, if a Molniya satellite is launched into an orbit that will carry it over Moscow

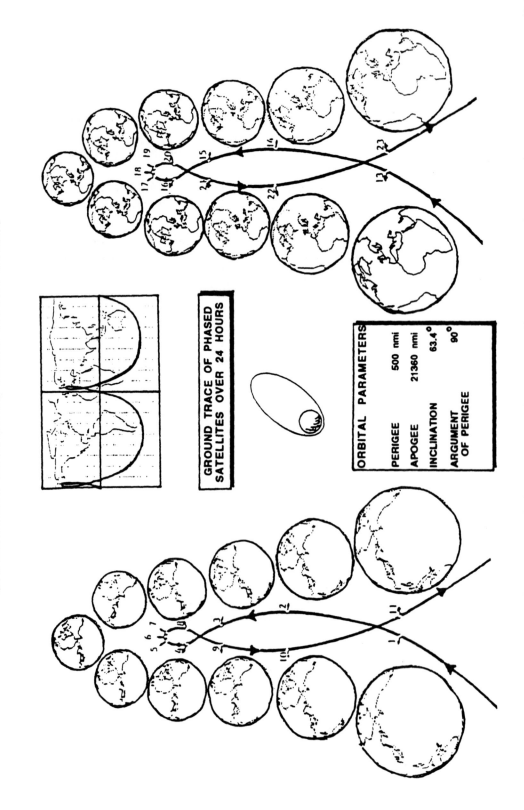

VIEWS OF EARTH FROM A SATELLITE IN A 12-HOUR CRITICALLY INCLINED ELLIPTICAL ORBIT

GROUND TRACE OF PHASED SATELLITES OVER 24 HOURS

ORBITAL PARAMETERS	
PERIGEE	500 nmi
APOGEE	21360 nmi
INCLINATION	63.4°
ARGUMENT OF PERIGEE	90°

COMMUNICATIONS POTENTIAL FOR SPECIAL
NONGEOSYNCHRONOUS ORBITS

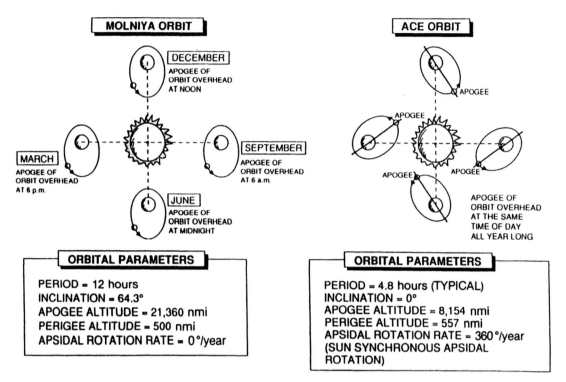

Figure 8.11 The apogee of a Molniya satellite always remains above its northernmost latitude, but, throughout the year, the satellite reaches its apogee location at different times of the day. The designers of the so-called ACE-orbit constellations have circumvented this difficulty by targeting their satellites for orbits in which the line of apsides twists at a constant rate of 360° per year. Consequently, the angle between the line of apsides and the sun vector always remains the same.

Figure 8.10 Rocketeers in the former Soviet Union have consistently favored inclined elliptical Molniya orbits for their communication satellites rather than the geosynchronous orbits usually employed by their counterparts in other spacefaring nations. This rather unusual choice is motivated by the fact that large numbers of Soviet citizens live at extremely high northern latitudes. Molniya satellites occupy elliptical orbits inclined 63.4° with respect to Earth's equator. That particular orbital inclination was selected because a satellite so inclined has a 0° apsidal rotation rate. Consequently, a Molniya satellite can keep its apogee always above its northernmost latitude.

in December at 12:00 noon, three months later in March, it will arrive at its apogee at 6:00 P.M. And three months after that, it will be over Moscow at 12:00 midnight. To help circumvent this problem, the Russian Molniya satellites are usually launched in groups of four working in partnership so that they can provide adequate coverage characteristics, with at least one of them hovering overhead at the right time during the different quadrants of the year.

The right-hand side of Figure 8.11 highlights an important property of a different kind of satellite that provides an alternative orbital geometry. These so-called ACE-orbit satellites[1] are always launched into elliptical orbits with 0° orbital inclinations. Their orbital elements are purposely chosen so the line of apsides will rotate exactly 360° per year. Thus, as the right-hand side of Figure 8.11 indicates, the angle between the Sun and the line of apsides always remains essentially constant as the seasons change.

Various orbital periods can be chosen for ACE-orbit satellites, but they are always rigged with repeating ground-trace geometry. This means that an ACE-orbit satellite completes an integral number of orbits in an integral number of days. The ACE orbit satellite depicted in Figure 8.11, for example, has an orbital period of 4.8 hours. Thus, it completes five orbits every day.

ACE-orbit satellites have special orbital geometries that can provide a number of advantageous characteristics. The orbital period and the ground-trace geometry of such a satellite, for example, can be chosen to serve specific, populated regions of Earth at particular times of the day. Specifically, a satellite in a particular ACE orbit might provide daily communication coverage over London, New York, and Bombay during the peak-load intervals for each of those major population centers. Thus, it could handle some of the peak-load communication traffic when the geosynchronous satellites serving those regions otherwise, might be overloaded.

Another advantage of the ACE-orbit satellites is that they do not require geosynchronous slot assignments. The apogee altitude of a typical ACE-orbit satellite is about one-third as high as a geosynchronous satellite. All ACE orbits are equatorial (0° inclination), so a thin beam of radio energy directed from a city above or below the equator toward an ACE-orbit satellite

---

[1] The acronym ACE stands for this rather awkward characterization: Apogee at Constant time-of-day Equatorial.

will not tend to illuminate the family of geosynchronous satellites in their higher-altitude orbits. The satellites launched into ACE orbits also have lower free-space losses because they are closer to their ground-based transmitters.

On the other hand, continuous satellite tracking is necessary because ACE-orbit satellites are not synchronous with Earth. The radiation environment also tends to be more damaging for ACE-orbit satellites, and they provide even worse coverage for extreme northern and southern latitudes than conventional geostationary satellites. However, despite these rather serious shortcomings, ACE-orbit satellites may eventually find a number of advantageous applications.

FROZEN ORBITS

Most satellites experience noticeable variations in orbital eccentricity and argument of perigee caused by the higher order potential harmonics (shape terms) of the earth's gravitational field. Fortunately, the distorting effects of these higher order harmonics can be induced to cancel one another provided the mission planners are free to choose the proper combination of initial *orbital altitude, inclination, eccentricity*, and *argument of perigee*. When choices of this type are possible, skilled experts can, for example, achieve nearly constant and persistent values for the *mean eccentricity* and the *argument of perigee* of a satellite's orbit.

A satellite whose orbital parameters are controlled by this technique is said to be in a "Frozen Orbit." Practical examples of satellite that have been launched into Frozen Orbits include:

- The Topex/Poseidon oceanographic satellite jointly sponsored by the Jet Propulsion Laboratory and the French government
- The Earth Radiation Budget Satellite used in studying the overall thermal properties of the earth
- The Landsat 4 and 5 Earth Resources Satellites
- Landsat 7 which is scheduled for launch in the second quarter of 1998

Depending on the mission requirements, specific Frozen Orbits can be selected to achieve a variety of practical results. These include satellites whose orbital altitude decays uniformly with atmospheric drag, satellites with essentially no ground-trace variations due to perigee (apsidal) rotation, and satellites with extremely small altitude variations over the Northern Hemisphere (or conversely over the Southern Hemisphere).

THREE-BODY, LIBRATION-POINT ORBITS

Generally speaking, the satellites discussed so far have been attracted primarily by the gravity from a single large celestial body such as Earth or the Moon. In some cases, however, a space vehicle that is purposely placed at the proper location in space may be attracted simultaneously by two or more powerful central force fields. When this happens, some interesting orbital characteristics and geometrical properties can result.

Consider Figure 8.12, for instance, which indicates that there are five stable locations in the vicinity of Earth and the Moon. A space vehicle launched into one of these so-called *libration-point orbits* will remain motionless with respect to those two strong gravitational fields. The libration-point orbits were discovered by the mathematician Joseph-Louis Lagrange in the eighteenth century. Notice that there are a total of five stable libration points at specific locations around Earth and the Moon. Three of them lie along the straight line connecting the two bodies; the other two occupy the equilateral points 240,000 miles from both Earth and the Moon.

The Moon, of course, is circling around the Earth–Moon barycenter with a period of about one month. As it makes its 2000 mph journey, the libration-point satellites are carried along with it in roughly circular "orbits." In other words, Earth, the Moon, and the five libration points always remain in a fixed configuration with respect to one another as the Moon and Earth whirl around their common barycenter.

One of the co-linear libration points (L_1) lies between Earth and the Moon, another (L_2) lies on the back side of the Moon as seen from Earth, and the third (L_3) lies on the back side of Earth as seen from the Moon.

The two remaining libration point orbits $(L_4$ and $L_5)$ are positioned at the equilateral points in the Earth–Moon plane. One leads Earth and the Moon through space, the other trails behind as the Moon orbits Earth. The equilateral points, of course, are 240,000 miles away from both Earth and the Moon.

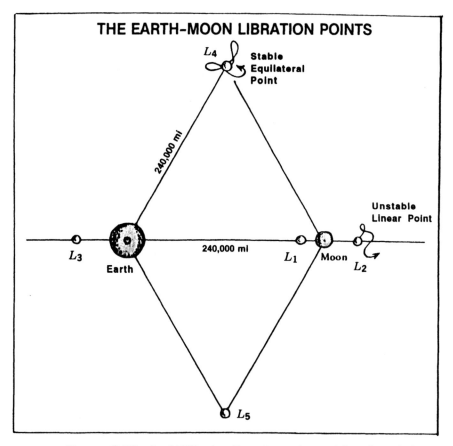

THE EARTH–MOON LIBRATION POINTS

L4 Stable Equilateral Point

240,000 mi

L3
Earth

240,000 mi

L1

Moon

Unstable Linear Point

L2

L5

Figure 8.12 In 1772, the French mathematician Joseph-Louis Lagrange discovered that there are five stable positions at which a satellite can be made to remain stationary with respect to Earth and the Moon. Three of these stable points lie along the rotating Earth-Moon radius vector. The other two are located at the equilateral points, in the Earth–Moon plane, equidistant from both the Earth and the Moon.

The equilateral libration points are dynamically stable. When a satellite positioned at one of the two equilateral libration points is subjected to a small force of perturbation, its orbit will be distorted, but it will remain in the vicinity of its libration point. The other three libration points are dynamically unstable. If a satellite in one of these three positions is subjected to a small force of perturbation, it will relentlessly drift away. With sufficiently small forces of perturbation, this process may take several months or years, but, eventually, the satellite will escape from its dynamically unstable location.

The five libration points exist because, at these particular locations, there is a permanent balance between the centrifugal force hurling the satellite away from the common barycenter and the gravitational forces pulling it toward both the larger and the smaller body. The vector sum of these three forces at each of the five libration points always equals zero.

Various practical applications have been proposed for the libration-point orbits, some of which have already been implemented. They can be, for example, used to monitor the potentially damaging particles traveling toward Earth from the Sun. Several years ago, the now defunct "L5 Society" proposed the construction of a large "glove" at the L_5 libration point to be used to "catch" materials hurled up from the surface of the Moon by electromagnetic catapults.

The naturally occurring Trojan asteroids are known to spiral around the Jupiter–Sun equilateral libration points. One of the "Great Telescopes," the Space Infrared Telescope Facility (SIRTF), was also, at one time, slated to occupy the L_2 libration-point orbit behind Earth as seen from the Sun.

In keeping with this plan, the supercooled SIRTF telescope was to be launched into a so-called "halo orbit." Halo orbits were assigned their unique name because they resemble the halo over the head of a Catholic saint. Once a satellite had been established in a proper halo orbit around L_1, it would travel along a roughly circular trajectory indefinitely, provided mission planners are willing to execute small, periodic corrective burns with its onboard rockets.

In the 1980s, a talented U.S. aerospace engineer named Robert Farquahr proposed the establishment of a communication link to be positioned in the L_2 halo orbit on the backside of the Moon. Farquahr's innovative concept called for a halo-orbit satellite circling around Earth–Moon L_2 libration point to provide continuous radio relays for space colonies on the Moon. With a corrective ΔV of only 400 feet per second per year, that relay satellite could be maintained in a halo orbit that would provide continuous communication connectivity between the lunar colony and various radio antennas located on Earth. With only 40 feet per second, Farquahr estimated that connectivity could be provided almost all the time.

INTERPLANETARY TRAJECTORIES

In 1925, the German researcher, Walter Hohmann, conjectured that the best way to send a space probe onto a Sun-centered interplanetary trajectory from Earth to Mars was to use a so-called Hohmann transfer maneuver. The Hohmann transfer maneuver, shown in Figure 4.3 of Chapter 4, is being used to carry a communication satellite from a 100-nautical-mile parking orbit to geosynchronous orbit. Notice that the maneuver sequence involves two separate impulsive rocket burns separated by a 180° coasting arc. Both velocity increments are added parallel to the existing velocity vector, so apparently no energy is wasted. Most direct interplanetary trajectories approximate the Hohmann transfer maneuver; however, their geometries and their timing are somewhat more complicated because the space probe must arrive when the planet is at the proper location. The planetary orbits are also slightly elliptical and they are inclined at slightly different angles with respect to one another.

Mars-mission opportunities involving reasonable ΔV's typically occur in pairs fairly close together. These paired opportunities come along approximately every two years. One of the opportunities calls for a fast transfer; the other calls for a slow transfer. On a fast transfer, the space probe travels less than 180° in reaching the orbit of Mars, and the transfer arc along which it travels to reach Mars is entirely internal to the orbit of Mars. On the slow transfer, the travel distance is more than 180° and the transfer orbit carries the probe beyond the orbit of Mars during a portion of its journey.

The total velocity increments required to fly directly from Earth to the two inner planets, Venus and Mercury, are presented at the bottom of Figure 8.13. The missions being considered do not involve any swing-by maneuvers. The total ΔV needed to reach Venus is approximately 37,000 feet per second. The total ΔV needed to reach Mercury lies somewhere between 41,500 to 47,500 feet per second depending on the precise geometry associated with the departure opportunity.

As Figure 8.13 indicates, traveling to within one-tenth of an astronomical unit of the Sun (approximately one million nautical miles) requires a velocity increment of 90,000 feet per second. However, most proposals for flying that close to the Sun employ energy-saving Jupiter swing-by maneuvers. In such a maneuver, the space probe starts by flying out to Jupiter where it employs a swing-by maneuver to bend its velocity vector back toward the Sun. This approach increases the total travel time, but it can save tens of thousands of feet per second.

The velocity increments needed to travel along direct injection trajectories (no swing-by maneuvers) from Earth to the

VELOCITY INCREMENTS NEEDED TO REACH
THE VARIOUS PLANETS

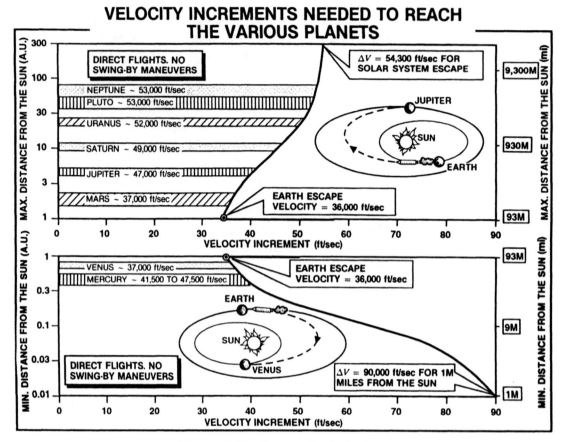

Figure 8.13 The total velocity increments needed to fly a space probe from a low-altitude Earth orbit to the various planets in the solar system are graphically depicted. A direct trip to Venus typically requires about 37,000 feet per second. A mission that carries a probe to Mercury requires somewhere between 41,500 to 47,500 feet per second. The outer planets require ΔV increments ranging from 37,000 to 53,000 feet per second. Curiously, our Sun is the most difficult star in the universe for earthings to reach at a ΔV cost of 90,000 feet per second. A ΔV of only 54,300 feet per second would allow a space probe to escape from our solar system without swingbys and coast to any other distant star.

exterior planets (including Jupiter) are presented at the top of Figure 8.13. The horizontal scale in this figure represents the total velocity increment, and the vertical scale denotes the maximum distance away from the Sun (in astronomical units). Any distant point in the universe can be reached with a velocity increment of 54,300 feet per second or less because such a velocity increment will allow the probe to escape the solar system and coast to any distant star. Strange as it may seem, our Sun is, by far, the most difficult star in the universe for us to reach in terms of total ΔV!

Reaching the planet Jupiter requires a total velocity increment of 47,000 feet per second. But, once the probe has reached Jupiter, a swing-by maneuver, properly timed and executed, can be used to reach almost any location in the solar system, or beyond. Reaching Saturn directly requires a velocity increment of 49,000 feet per second. Uranus, Neptune, and Pluto require total ΔV's ranging between 52,000 and 53,000 feet per second.

GRAND TOUR MISSIONS

Researchers at Cal Tech discovered that during 1977, the planets Jupiter, Saturn, Uranus, and Neptune would achieve an extraordinary alignment such that three successive swing-by maneuvers could be used to send a space probe on a flyby visit to all four of those exterior planets for the price of one. This so-called "Grand Tour Mission" caused a flurry of excitement among U.S. mission planners and aerospace engineers. On the Grand Tour Mission, a multistage rocket launched from Earth would send a space vehicle onto a swing-by trajectory around Jupiter. After executing that precisely controlled swing-by maneuver, the probe would arrive at Saturn and again, swing by Saturn to arrive at Uranus where one final swing-by would carry it out to Neptune.

As Cal Tech researcher Homer Stewart demonstrated, the planets come into the proper alignment for the Grand Tour Mission only once every 175 years. Two consecutive Grand Tour opportunities occurred during Thomas Jefferson's presidency and again in 1977. The next opportunity will occur in 2152.

Practical swing-by missions sometimes take considerably longer than corresponding direct flights, but the Grand Tour Mission provides important reductions in both total ΔV and total mission time. Four planets could be visited for the price of one (in terms of ΔV) but, in addition, the total travel time turned out to be only about 9 years, compared with 30.9 years for an equivalent set of direct-flight missions.

In the 1970s, NASA managers attempted to get the U.S. Congress to approve the Grand Tour Mission as a formal line item in the federal budget. The mission never achieved that exalted status, but two Voyager space probes did fly along Grand Tour swing-by trajectories with departures from Earth orbit in 1977. Voyager 2 (see Figure 8.14) swooped by Jupiter, Saturn, Uranus, and Neptune. Then it was jerked onto an escape trajectory to exit the solar system along a hyperbola parallel to the ecliptic plane. The Voyager 2 spacecraft encountered Jupiter and Saturn, and then in a final swing-by it was jerked onto an escape trajectory perpendicular to the ecliptic plane.

Both Voyager space probes carried special engraved plaques specifically designed to yield important information about us to any intelligent extraterrestrial creatures who might run across them in the vicinity of some distant star.

SPECIAL ORBITS IN REVIEW

Figure 8.15 summarizes some of the beneficial properties associated with the special orbital locations surrounding Earth. The radial parameter in this polar-coordinate graph corresponds to the orbital radius and the angular parameter represents orbital inclination. The three-dimensional movement of each satellite around Earth is not shown.

The cluster of circles positioned above Earth's equator at 6.6 Earth radii represents the large family of geosynchronous satellites in 0° (equatorial) orbits. As the note above the North Pole in the figure indicates, a satellite coasting around a polar orbit at any altitude will have a nodal regression rate equal to zero; its inertial nodal crossing point will not be systematically perturbed by Earth's equatorial bulge.

The circular Sun-synchronous orbits all lie along the curving arc near the center of the polar-coordinate graph. Satellites launched into dawn–dusk circular orbits anywhere in the altitude range lying between 750 and 1800 nautical miles will remain in full sunlight all the time.

Figure 8.14 In 1977, two Voyager space probes were launched on "Grand Tour" Missions to the planets in the outer solar system. Voyager 1 traveled by Jupiter and Saturn, where a second swing-by maneuver jerked it up to escape velocity perpendicular to the ecliptic plane. Voyager 2 flew by Jupiter, Saturn, Uranus, and Neptune, and then it, like Voyager 1, was hurled onto an escape parabola that is carrying it along a gently curving trajectory deep into the universe.

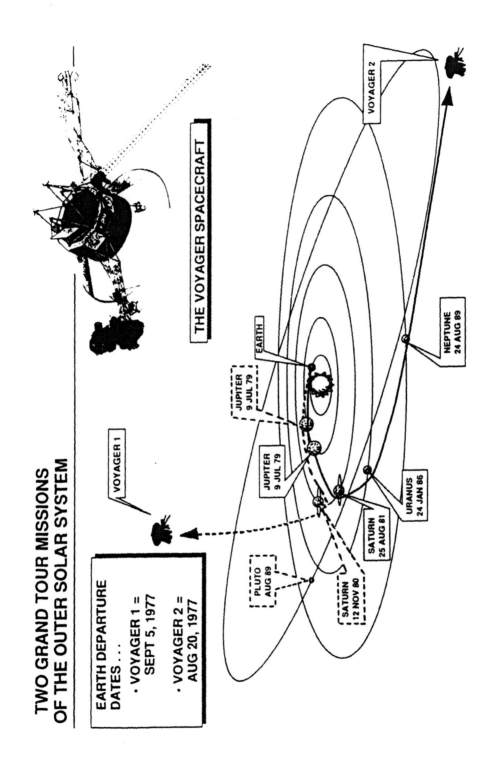

TWO GRAND TOUR MISSIONS
OF THE OUTER SOLAR SYSTEM

EARTH DEPARTURE DATES....
- VOYAGER 1 = SEPT 5, 1977
- VOYAGER 2 = AUG 20, 1977

VOYAGER 1

THE VOYAGER SPACECRAFT

VOYAGER 2

EARTH

JUPITER
9 JUL 79

JUPITER
9 JUL 79

SATURN
25 AUG 81

SATURN
12 NOV 80

PLUTO
AUG 89

URANUS
24 JAN 86

NEPTUNE
24 AUG 89

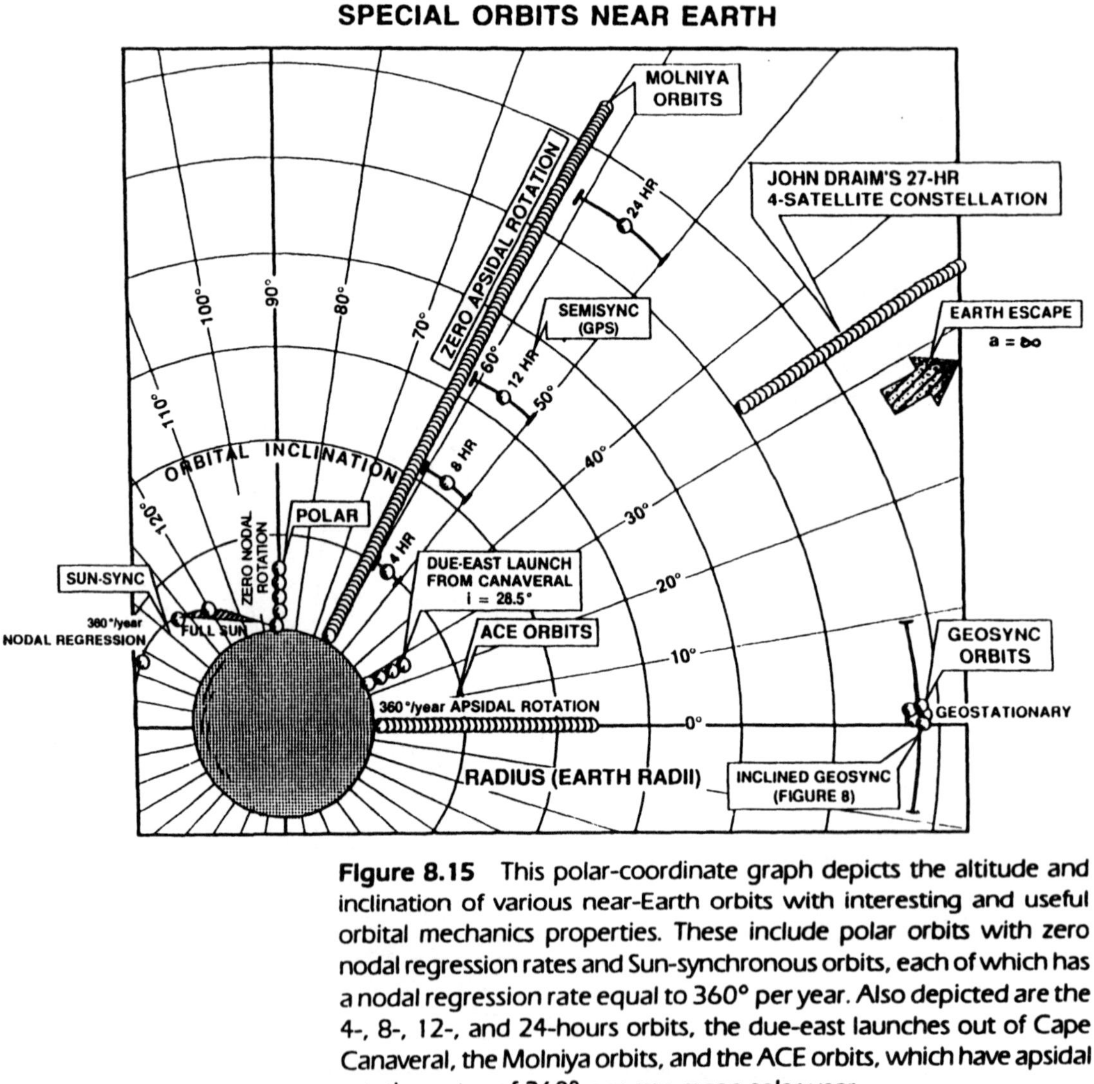

SPECIAL ORBITS NEAR EARTH

Figure 8.15 This polar-coordinate graph depicts the altitude and inclination of various near-Earth orbits with interesting and useful orbital mechanics properties. These include polar orbits with zero nodal regression rates and Sun-synchronous orbits, each of which has a nodal regression rate equal to 360° per year. Also depicted are the 4-, 8-, 12-, and 24-hours orbits, the due-east launches out of Cape Canaveral, the Molniya orbits, and the ACE orbits, which have apsidal rotation rates of 360° per one mean solar year.

The 12-hour Russian Molniya orbits are also depicted in Figure 8.15. Notice that they have inclinations of 63.4°. Any satellite placed in an elliptical orbit with that particular inclination will have a zero apsidal rotation rate. The Russian Molniya satellites are launched into highly elliptical 12-hour orbits with low-altitude perigees and apogees that are slightly beyond the geosynchronous altitude. This unique orbital geometry is indicated by the overlapping circles lying along the zero apsidal rotation line (63.4°). The Russian Molniya satellites are rigged so that their apogees always remain above the northernmost latitude.

The orbital altitudes of satellites with periods of 4, 8, 12, and 24 hours are shown just to the right of the zero apsidal rotation line. For convenience, their symbols are positioned at 55° inclinations. However, a satellite with any desired orbital period can be launched into any orbital inclination. This is indicated symbolically by the short arcs running through their ball-like symbols.

Low-altitude satellites launched out of Cape Canaveral are represented by the circles running along the 28.5° inclination line. These are due-east launches that obtain the maximum possible free velocity from Earth's rotation. The special ACE-orbit satellites discussed earlier in this chapter are also represented in Figure 8.15. Notice that they are launched into elliptical equatorial orbits. The line of apsides for an ACE orbit rotates 360° per mean solar year. Consequently, the angle between its line of apsides and the radius vector to the Sun remains essentially constant throughout the year.

Figure 8.16 highlights the special orbits associated with the Earth–Moon system including the figure-8 free-return trajectory, the equal-period descent orbit, and the various Earth-Moon libration-point orbits.

Notice how a free-return figure-8 trajectory loops around the Moon before it curls back to the vicinity of Earth. This special orbit provided an extra margin of safety for the Apollo astronauts when they flew from Earth to the Moon. If, for any reason, they had been unable to execute their preplanned braking maneuver in the vicinity of the Moon, their spacecraft would have automatically coasted back into the vicinity of Earth.

Figure 8.16 also depicts the equal-period descent maneuver, which the astronauts used when they traveled down toward the surface of the Moon. Once they were in the equal-period descent orbit, they could have skipped the landing sequence, if necessary, and still ended up on an automatic rendezvous orbit with the parent spacecraft circling overhead. An equal-period descent maneuver is executed by making a short-duration rocket burn radially toward or radially away from the center of the Moon. This special energy-wasting maneuver changes the direction of the vehicle's velocity vector without changing its magnitude. Consequently, the semi-major axis and the orbital period of the spacecraft also remain unchanged.

Once they were traveling along an equal-period descent orbit, the astronauts could descend down toward the Moon's surface with confidence. If they were unable to find a safe landing spot, they could have coasted past perigee and auto-

SPECIAL ORBITS FOR THE EARTH-MOON SYSTEM

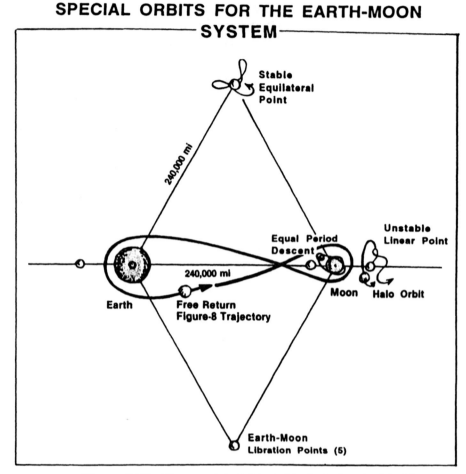

Figure 8.16 The special orbits and trajectories depicted here are affected by the gravitational attractions from both Earth and the Moon. The so-called free-return figure-8 trajectory carried the Apollo astronauts along their three-day cis-lunar journey. The equal period descent maneuver was then used during their descent to the lunar surface because it provided an automatic return to their parent spacecraft, circling overhead. The three co-linear libration-point orbits and the two equilateral libration-point orbits are also depicted. They were discovered mathematically by the French mathematician Joseph-Louis Lagrange in the eighteenth Century.

matically returned into the vicinity of the Apollo capsule circling overhead where they could have been rescued for a safe return back toward Earth.

The five Earth–Moon libration points are also highlighted in Figure 8.16. The three co-linear libration points lying along the Earth–Moon line are dynamically unstable. The other two, which lie at the empty apexes of the two equilateral triangles, are dynamically stable. One of the so-called halo orbits is also depicted in the figure. A satellite placed in a halo orbit around the L_2 libration point theoretically could provide a continuous communication relay link to serve a colony of lunar explorers situated on the backside of the Moon.

When satellite designers are planning to launch a satellite or a space probe, they should consider various possibilities for taking advantage of the special geometrical properties associated with the various special orbits in the Earth–Moon system. These include Sun-synchronous orbits, geosynchronous and Molniya orbits, ACE orbits, and 28.5° due-east orbits. This analysis approach can be further refined when they are planning to launch a complete constellation of satellites—an intriguing subject that is discussed in detail in Chapter 9.

9 CHOOSING THE PROPER CONSTELLATION ARCHITECTURE

Newton's peculiar gift was the power of holding continuously in his mind a purely mental problem until he had seen through it. —*Economist John Maynard Keynes* (Essays in Biography, *1933*)

Previous chapters have included discussions of some of the methods engineers use in selecting the altitude, inclination, and various other defining parameters associated with the orbit of a single satellite. This chapter deals with the rationale used in selecting the characteristics of an entire constellation.

Generally speaking, we choose to launch a constellation of satellites for only one reason: to provide some specific type and level of coverage. The coverage we are seeking may be, for example, continuous global coverage, intermittent global coverage, or continuous local coverage. Multiple satellite coverage is also a common goal of today's constellation designers and aerospace engineers. Those of us who masterminded the Navstar GPS constellation, for instance, needed to provide at least quadruply-redundant coverage to millions of users around the globe. This rather demanding requirement came about because widely dispersed navigators need continuous access to at least four satellites to obtain the accurate three-dimensional navigation solutions they desired.

As we have seen, some constellations are designed to cover only a portion of Earth. The Russian Molniya satellites, for instance, are usually launched in batches of 4 into orbits carefully structured to provide continuous around-the-clock coverage for the vast Soviet territory, all of which is located in the Northern Hemisphere.

WHAT IS A CONSTELLATION?

A constellation can be defined as *a collection of satellites of similar design performing similar functions in similar orbits at the same time*. This definition is reasonably accurate, but it is difficult to remember. Another way to gain insight into the basic concept of a constellation is to examine a few typical examples from the real world. Knowledgeable professionals would probably agree that various collections of satellites such as Transit, Iridium, GPS, and the Soviet Glonass are definitely constellations, and so is the Russian family of Molniya satellites.

A Sampling of Today's Constellations

Table 9.1 describes the key features and the selection rationale for a sampling of today's constellations. The Navstar GPS, for example, which is the first constellation listed in the table, was specifically designed to provide continuous ranging measurements with at least four satellites above the horizon at all times everywhere in the world. The GPS constellation consists of 24 satellites in semisynchronous (12-hour) orbits launched into six different orbital planes each tipped 55° with respect to the equator.

Mission planners selected that particular constellation because they knew it would provide continuous, global, four-satellite visibility to the users with the satellites widely scattered across the sky. The military sponsors of the GPS constellation also wanted to be able to upload each satellite at frequent and predictable intervals from the continental United States. This special requirement led us to select a 12-hour orbital period because, when the satellites are launched into circular 12-hour orbits, they appear repeatedly over the same point on Earth approximately every 24 hours.

The Transit constellation, which is also devoted to globe-spanning navigation, was selected by using a completely different rationale. The Transit navigation satellites are designed to provide global, but intermittent, radio navigation coverage using Doppler-shift measurements from each individual satellite as it sweeps across the sky. This crude navigation technique led the designers to position the Transit satellites in six polar birdcage orbits 580 nautical miles above Earth. The low-altitude location for the Transit satellites was chosen, in part, to achieve rapidly varying Doppler-shift variations for users on the ground.

TABLE 9.1 Typical Existing Constellations

Constellation	Purpose	No. of Satellites (No. of Orbit Planes)	Altitude (Inclination)	Orbit-Selection Rationale
Navstar GPS	Continuous global radio-navigation using pseudo-ranging to at least 4 satellites	24 semisync, 12-hour orbits (six planes)	10,898 nmi (55°; 63° for Block I's)	• Continuous, global, 4-satellite visibility with good GDOP • Uploads at predictable intervals from U.S.-controlled territory
Soviet Glonass	Continuous global radio-navigation using pseudo-ranging to at least 4 satellites	24 subsemisync (three planes)	10,313 nmi (64.8°)	• Continuous, global, 4-satellite visibility with good GDOP • Uploads from Soviet territory • GPS navigation compatibility?
Transit navigation satellite	Intermittent global radio-navigation using Doppler-shift measurements	6 low-altitude, polar orbits (six planes)	580 nmi (90°)	• Global, but intermittent coverage • Rapidly varying Doppler shift
Soviet Cicada	Intermittent global radio-navigation using Doppler-shift measurements	6 low-altitude, polar orbits (six planes)	Low altitude (90°)	• Global, but intermittent coverage • Rapidly varying Doppler shift • Transit navigation compatibility?
Intelsat communication satellites	Active global communications relay	several geosync, equatorial orbits (one plane)	19,300 nmi (0°)	• Essentially global continuous coverage • No antenna steering for ground antennas. Small range variations

(Continued)

TABLE 9.1 *Continued*

Constellation	Purpose	No. of Satellites (No. of Orbit Planes)	Altitude (Inclination)	Orbit-Selection Rationale
Soviet Molniya	Active communications relay for all Soviet territory	4 elongated, elliptical orbits (two planes)	Perigee = 300 nmi Apogee = 20,000 nmi (63.4°)	• High inclinations to cover northern latitudes • Elliptical to spend most time in Northern Hemisphere • 63.4° inclinations to eliminate apsidal rotations
Landsat constellation	False-color images for earth resources studies worldwide	2 low-altitude, Sun-sync orbits (two planes)	380 nmi (~98°)	• Low altitude for good viewing of Earth resources • Sun synchronous to achieve good viewing with slanting rays from the Sun

The Intelsat constellation is designed to provide real-time, line-of-sight communication relays for a substantial fraction of the globe. This special requirement motivated the constellation designers to place their satellites in geostationary orbits 19,300 nautical miles high. The resulting constellation provides essentially continuous global coverage, except for two small, largely uninhabited, regions surrounding the poles. Geostationary satellites also allow minimal steering for the ground antennas with very small range variations between the users and the satellites.

The constellation architecture for the Russian Molniya satellites was selected with a different service characteristic in mind: The former Soviet Union has large numbers of potential users living at high northern latitudes. Geosynchronous communication satellites provide poor coverage at these extreme latitudes because the satellites tend to be situated near the horizon. Consequently, the Russian engineers decided to launch their satellites into highly elongated elliptical orbits with 12-hour orbital periods inclined 63.4° with respect to the equator. That particular orbital inclination was chosen to eliminate apsidal rotations that would have eventually carried the apogee of the satellite's orbit over the Southern Hemisphere. The elongated orbits the Russian engineers selected cause the Molniya satellites to spend most of their time at high northern latitudes where they can serve large numbers of residents of the former Soviet Union.

The Landsat constellation (see Table 9.1) was designed to provide high-resolution, false-color images for Earth resources studies conducted on a global scale. Those who engineered the Landsat selected a constellation with two satellites occupying 380-nautical-mile, Sun-synchronous orbits tipped 98° with respect to the equator. Low-altitude, Sun-synchronous orbits provide excellent viewing for earth resources observations. Sun-synchronous orbits with repeating ground-trace geometry were selected to provide repeatable lighting conditions whenever the satellite was above the same regions on Earth.

Spatial Locations for the Selected Constellations

The constellation architectures for the six families of satellites discussed so far are depicted graphically in Figure 9.1. The radial variable in this polar-coordinate graph represents the distance between the center of Earth and the satellite. The angular coordinate represents the orbital inclination of the various satellites in the constellation.

CONSTELLATION ARCHITECTURES

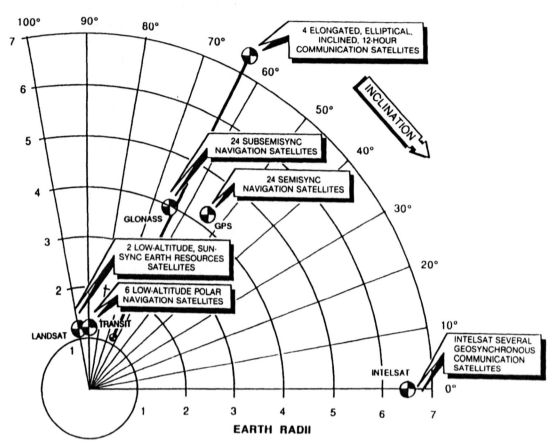

Figure 9.1 The radial variable in this polar-coordinate graph represents the distance between the center of Earth and a satellite. The angular coordinate represents the satellite's orbital inclination. The six constellations depicted range from near-polar, low-altitude, Sunsynchronous satellites to geosynchronous satellites positioned 19,300 nautical miles above Earth.

Notice how the Intelsat communication satellites are positioned at a distance of 6.6 Earth radii from the center of Earth. All of them are represented by a single round symbol in this special graphical depiction, although they are positioned, of course, at various locations above the equator. The six Transit navigation satellites are launched into low-altitude polar orbits. Consequently, their icons are all located near the center of the graph. The Landsat Earth resources satellites are positioned nearby in low-altitude, Sun-synchronous orbits. The Navstar GPS satellites are lofted into 55° orbital inclinations about 4.16 Earth radii (10,898 nautical miles) from the center of Earth. The Russian Glonass satellites have slightly higher inclinations (64.8°) but they are about 600-nautical-miles lower than the GPS satellites. At that lower altitude, each Glonass satellite completes 17 orbits every eight days. The Russians launch their Molniya constellations into highly elongated elliptical orbits with 63.4° inclinations. At apogee, which is positioned at the northernmost latitude, the Molniya satellites are slightly above the geosynchronous altitude.

As Table 9.1 indicates, constellations of satellites are often placed in orbits with special geometrical properties. These special orbits include polar, Sun-synchronous, geosynchronous, semisynchronous, and Soviet Molniya orbits. Working together in partnership, these special-orbit constellations provide abundant coverage for their clients most of whom are situated on the ground below.

WHAT IS THE LARGEST CONSTELLATION OF SATELLITES EVER LAUNCHED INTO SPACE?

In my three-day short courses titled "Launch Vehicles and Orbital Mechanics," I sometimes ask my students this seemingly simple question: "What is the largest constellation of satellites ever launched into space?" The most frequent answer is: "The Navstar Global Positioning System," which consists of 24 satellites. Occasionally, one of my students will argue that the 7000 pieces of space debris tracked by the U.S. Air Force constitute a constellation of satellites. Both of these thoughtful answers are, however, off the mark by several orders of magnitude. In my view, the largest constellation of satellites ever launched into space was the 1.2 billion copper dipole needles that were the payload for Project West Ford. As Figure 9.2 illustrates, the West Ford needles were used to reflect faint radio messages from one Earth station to another at a frequency of 8 gigahertz.

Each copper dipole was three-fourths of an inch long and 1/500th of an inch in diameter. They were boosted into a 2000-nautical-mile orbit with an 87° inclination. Each needle weighed only 1/20 millionth of a pound, but, taken together, they tipped the scale at 50 pounds.

Two West Ford launches took place in the 1960s. In the first one, the deployment mechanism malfunctioned and spewed out the needles in useless clumps. In the second mission, they were successfully deployed into a donut-shaped belt of in-

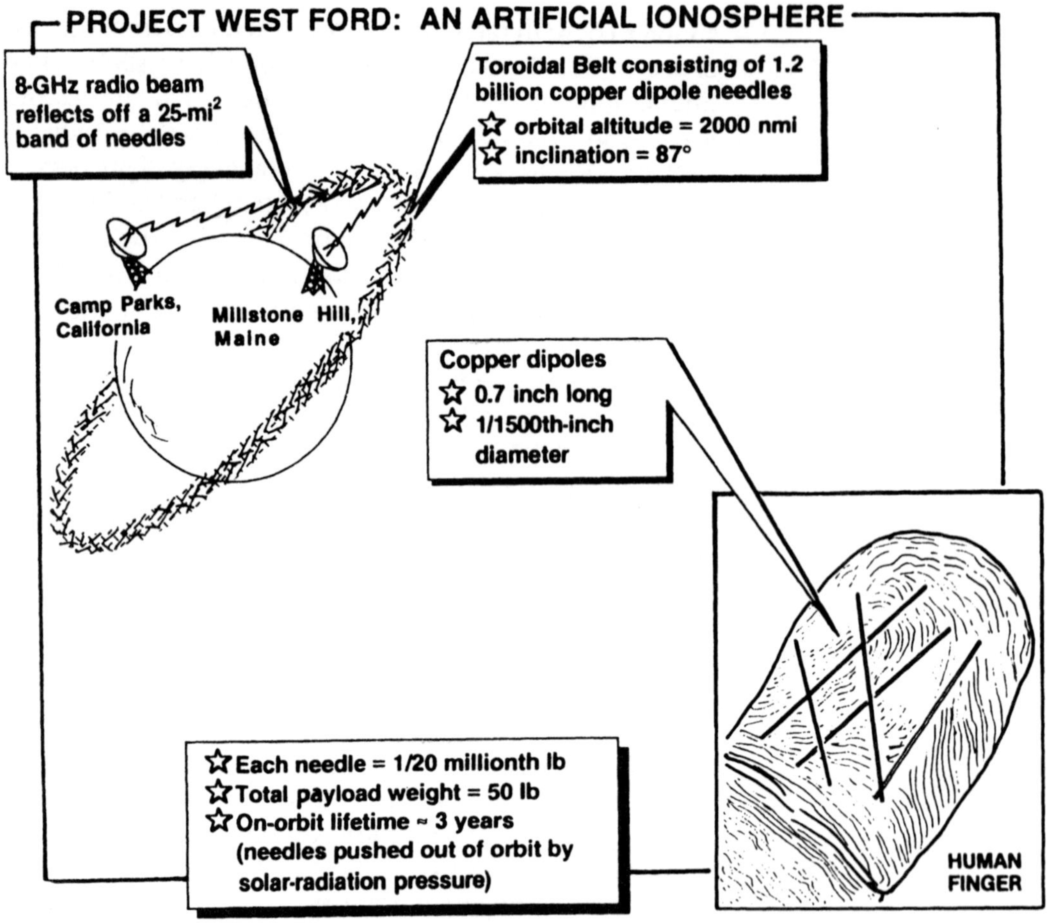

PROJECT WEST FORD: AN ARTIFICIAL IONOSPHERE

8-GHz radio beam reflects off a 25-mi^2 band of needles

Toroidal Belt consisting of 1.2 billion copper dipole needles
☆ orbital altitude = 2000 nmi
☆ inclination = 87°

Camp Parks, California   Millstone Hill, Maine

Copper dipoles
☆ 0.7 inch long
☆ 1/1500th-inch diameter

☆ Each needle = 1/20 millionth lb
☆ Total payload weight = 50 lb
☆ On-orbit lifetime ≈ 3 years
(needles pushed out of orbit by solar-radiation pressure)

HUMAN FINGER

Figure 9.2 The largest constellation of satellites ever launched into space was Project West Ford, which consisted of 1.2 billion copper dipole needles each of which was three-quarters of an inch long and 1/500th of an inch in diameter. The needles formed a donut-shaped belt around Earth used in bouncing radio messages from one terrestrial location to another in the 8-gigahertz portion of the frequency spectrum.

dividual dipoles encircling Earth. They remained in space about three years before solar-radiation pressure gently pushed them into Earth's atmosphere where aerodynamic heating caused them to disintegrate into a harmless vapor.

WHAT IS THE SMALLEST CONSTELLATION OF SATELLITES THAT CAN COVER EARTH?

In the early 1960s when space exploration was still in its infancy, a number of aerospace experts concluded independently that six satellites was the smallest number that could provide continuous line-of-sight coverage for the globe. It is relatively easy to demonstrate that three sufficiently high satellites in circular polar orbits working in partnership with three more in circular equatorial orbits could provide continuous global coverage. However, a smaller number of satellites will suffice.

John Walker's Rosette Constellations

John Walker, an aerospace engineer in England, looked at the problem from a fresh perspective and correctly concluded that it was possible to provide continuous global coverage with a constellation consisting of only five satellites. Walker's constellations were constructed with satellites in inclined circular orbits with their nodal crossing points equally spaced around the equator. He called his compact swarms of satellites "Rosette Constellations." He selected that colorful and informative name because, when he drew his constellations as seen from a vantage point high above the North Pole, his sketches looked like little roses (rosettes).

John Walker demonstrated that a properly inclined constellation of five satellites positioned in circular orbits at any altitude above 6200 nautical miles can provide continuous coverage for the entire globe.

The graph in Figure 9.3 shows how many satellites are required to achieve continuous coverage in other orbital-altitude regimes. At the lower altitudes, of course, each satellite has line-of-sight access to a smaller fraction of the globe so large numbers of them are required. At an altitude of 1000 nautical miles, for instance, a 21-satellite constellation is needed for continuous global coverage.

John Draim's Four-Satellite Constellations

For many years, John Walker's five-satellite rosettes were regarded as the smallest constellations that could provide continuous global coverage. Later, however, John Draim, an American aerospace engineer, figured out how to do it with only four. Draim, who holds a patent on his special constellation, abandoned circular orbits, and decided, instead, to place his satellites in inclined *elliptical* orbits. Strictly speaking, continuous coverage is possible only if the four satellites are place in orbits with 27-hour periods or longer. However, John Draim found that if he reduced the average orbital altitude of his

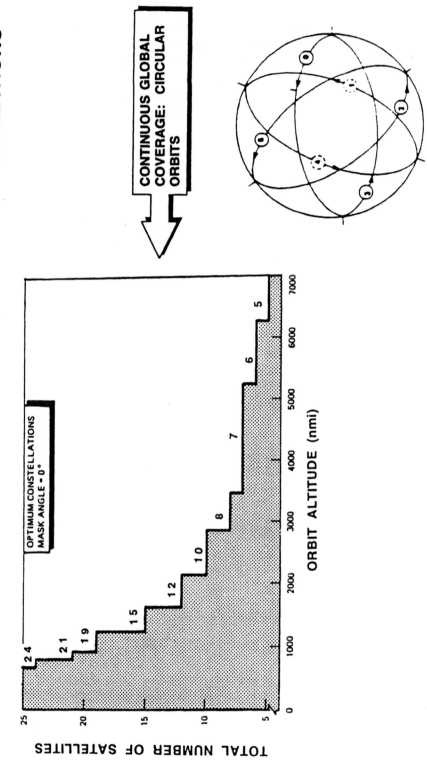

CONSTELLATION COVERAGE FOR WALKER CONSTELLATIONS

CONTINUOUS GLOBAL COVERAGE: CIRCULAR ORBITS

OPTIMUM CONSTELLATIONS
MASK ANGLE = 0°

TOTAL NUMBER OF SATELLITES

ORBIT ALTITUDE (nmi)

212

satellites to the 24-hour level, a properly designed four-satellite constellation would still provide 99.996 percent continuous coverage.

As Figure 9.4 indicates, John Draim's satellites are positioned in four different 31.3° inclined elliptical orbits, all of which have the same sizes and shapes. The semi-major axis of each 24-hour, Draim-constellation satellite is 22,767 nautical miles. If they are launched out of Cape Canaveral, Draim-constellation satellites could be 25 percent heavier than geosynchronous satellites riding aboard the same boosters. This is true primarily because of their much smaller plane-change requirements associated with Draim's inclined elliptical orbits.

Comparisons between the constellations of John Walker and those devised by John Draim are presented in the lower left-hand corner of Figure 9.4. Notice that for single, double or triple coverage, John Draim's constellations always require exactly one satellite less. Double coverage requires seven Walker-constellation satellites, for example, but only six satellites if John Draim's elliptical orbits are used instead.

The ground-trace geometry and the coverage characteristics for John Draim's 24-hour constellation are conceptually simple. Each of the two pairs of satellites traces out the same "circular" ground trace repeatedly. A careful study of the coverage patterns of the satellites indicates that, for all practical purposes, every portion of Earth has continuous, direct line-of-sight access to at least one satellite.

The Smallest Constellation That Can Provide Global Coverage

What is the smallest constellation of satellites that can cover the globe? The answer to that seemingly simple question is strongly dependent on the constraints we choose to apply. If

Figure 9.3 In the 1960s, the English aerospace engineer John Walker demonstrated that continuous global coverage could be achieved by a constellation of only five satellites, provided they were placed in the proper set of inclined circular orbits. This graph defines the number of satellites needed for continuous coverage at various orbital altitudes. At any altitude above 6200 nautical miles, five satellites arranged in a Walker "rosette" will suffice for continuous global coverage. At lower altitudes, larger numbers of satellites are required. At 1000 nautical miles, for instance, approximately 21 satellites properly positioned will provide the desired coverage characteristics.

JOHN DRAIM'S 4-SATELLITE
CONTINUOUS COVERAGE CONSTELLATION

NEAR-SYNCHRONOUS ORBITS

- INCLINED ELLIPTICAL NEAR-GEOSYNC ORBITS (31° INCLINATIONS)
- TWO CIRCULAR GROUND TRACES
- 4 SATELLITES PROVIDE 99.996% CONTINUOUS COVERAGE

PERFORMANCE ENHANCEMENTS

- SLOW, LUMBERING SATELLITE MOVEMENT
- INSIGNIFICANT RANGE VARIATIONS
- 25% PAYLOAD-WEIGHT GAIN OVER GEOSYNC SATELLITES: SMALL OR NO PLANE CHANGE REQUIRED

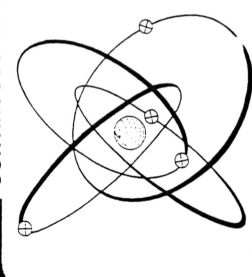

COVERAGE REQUIRED	NO. OF WALKER SATELLITES (CIRCULAR ORBITS)	NO. OF DRAIM SATELLITES (ELLIPTICAL ORBITS)
SINGLE-SATELLITE CONTINUOUS COVERAGE	5	4
DOUBLE-SATELLITE CONTINUOUS COVERAGE	7	6
TRIPLE-SATELLITE CONTINUOUS COVERAGE	9	8

we require that the satellites be placed in circular equatorial and polar orbits, the answer is *six*. Three of them would be launched into polar orbits with the other three in equatorial orbits.

If we allow the satellites to be placed in *inclined* circular orbits, we can get by with five satellites arranged in one of the rosette constellations designed by Englishman John Walker. If, however, elliptical orbits are permissible, we can achieve continuous global coverage with only *four* satellites using the special constellation devised by John Draim.

Of course, if we are willing to relax one more mission constraint and settle for *intermittent* global coverage rather than continuous coverage, a single satellite will suffice. We merely launch that single satellite into a properly selected orbit with repeating ground-trace geometry.

For a long time, both my students and I were convinced that one satellite was the minimum number required for global coverage, but then one of the participants in a Washington, D.C., session pointed out that we could achieve global coverage with *part of a satellite*. As he correctly concluded, this can be accomplished by leasing a transponder on any low-altitude, global-spanning satellite.

Almost immediately after that clever solution was proposed, another participant came up with a way for getting the job done with an even smaller number of satellites. We could, cover the globe he pointed out, by reflecting radar beams off the Moon or the ionosphere so we can get the job done with *zero satellites*!

COMPUTER-MODELING TECHNIQUES

A variety of computer-simulation programs have been developed to simulate the coverage characteristics of proposed satellite constellations. Two important questions must be

Figure 9.4 John Draim's four-satellite constellation consists of four satellites in inclined elliptical orbits with identical sizes and shapes. For his 24-hour constellation, which provides 99.996 percent continuous global coverage, the orbital inclination is 31.3°, the semi-major axis is 22,767 miles, and the orbital eccentricity is 0.2338.

answered before an effective computer-simulation program can be developed:

1. What kind of map projection provides the best visualization?
2. What kind of coverage analysis should we choose?

These two questions are addressed in what follows.

Map Projections

The orthographic projection, which is essentially a spherical representation of Earth, provides one popular and convenient map. The plastic globes found in elementary classrooms provide us with a common example of an orthographic projection. At Rockwell International, my friend Janis Indrikis and I used both a classroom globe and a spherical chalkboard as orthographic aids to help in the visualization of ground-trace projections and the coverage characteristics of satellites and their constellations.

Mercator projections are also commonly found in elementary schools. Those who make maps of this type project the surface of Earth onto a flat rectangle covered with a checkerboard pattern of longitude–latitude coordinates. The Mercator projection is easy to understand and familiar to nearly everyone. Unfortunately, it has one important disadvantage: areas are not preserved. If we draw a small circle of fixed size at different locations on a Mercator projection, the number of square miles of territory it will cover varies, depending on how far it is from the equator. Alaska is a big state. But it is artificially enlarged on a Mercator projection.

The Goode equal-area projection, which resembles a poorly peeled orange, preserves areas, but because of its oddly shaped contours, a Goode map does not provide a very convenient format for displaying the coverage characteristics of satellite constellations.

The Mollweide equal-area projection, which has a shape resembling an egg, represents a convenient compromise between the Goode equal-area projection and the more popular Mercator projection. The Mollweide equal-area projection is becoming increasing popular for use in connection with constellation coverage studies because it has so many advantageous characteristics.

Coverage Analysis

The second important choice aerospace engineers must make when they are developing a coverage routine is the fundamental methodology to be used in determining the constellation's coverage characteristics.

When he devised his four-satellite, elliptical-orbit constellation, John Draim relied on a simple mathematical theorem. He started by proving that if the satellites in a four-satellite constellation are connected to make a four-sided polyhedron, which entirely encloses Earth, that constellation will, at that particular instant in time, provide global coverage. Of course, the shape of the polyhedron changes as the satellites in the constellation move to new locations in the sky. Consequently, continuous global coverage is achieved only if Earth is always contained within the constantly evolving polyhedron.

Another popular method for determining the coverage characteristics of a satellite constellation is called a *grid search*. When a grid-search procedure is being used, Earth is divided into a checkerboard pattern sliced by parallel longitude-latitude grid lines. At each time point, the program simulates the locations of the various satellites, then it moves to the center of the first grid cell, "counts" the number of satellites that are visible from that location, and then automatically prints the resulting number in the center of the grid. If, for example, five satellites are visible from that location, the computer will print a "5" in the center of the cell.

It then moves to the next adjacent cell and carries out the same procedure. This process is repeated over and over again until the coverage characteristics of the constellation have been determined for the entire globe. When this global process has been completed, the computer moves the satellites forward to their next locations in accordance with the laws of orbital mechanics and then it performs a similar sequence of grid calculations.

Area calculations are another method for ascertaining the coverage characteristics of a constellation of satellites. In this case, the viewing regions (footprints) for a set of orbiting satellites are laid down by the computer one by one as contours on some convenient map projection. Then its computational routines automatically determine what fraction of Earth is covered by different numbers of satellites.

THE SPACE EGGS
COMPUTER-SIMULATION PROGRAM

A superbly designed computer-simulation program developed by Janis Indrikis and Bob Cleeve makes clever use of the Mollweide equal-area map projection and the pixel-manipulation capabilities of modern microcomputers. It also takes full advantage of their color-graphics capabilities. Figure 9.5 summarizes the salient characteristics of this computer program, which is called the "Space Eggs" simulation. The Space Eggs simulation program begins by displaying a Mollweide equal-area projection onto the computer screen. For a typical computer display, the resulting world map covers approximately 100,000 pixels. The color of each pixel is represented by a multidigit binary number.

The coverage regions of the various satellites are dropped onto the screen one at a time and the corresponding pixels are automatically incremented by one unit in binary notation. This causes the screen colors to change automatically in response. The pixels with a zero count are colorless. Those that have a pixel count of one might be red, for instance. Those that have a pixel count of two might be blue, and so on.

When all of the coverage contours have been dropped onto the screen, the computer carries out a series of pixel counts to determine what fraction of Earth is covered by what number of satellites. When a Vax minicomputer is being used, no pixel counts at all are required. The Vax minicomputer already "knows" the appropriate pixel counts so they can be displayed immediately in a convenient pictorial format.

The Space Eggs computer-simulation program automatically updates the location of the satellites as they sweep across the sky. At each new time point, it constructs a color map showing the number of satellites visible at each position around the globe.

Figure 9.5 The "Space Eggs" computer-simulation program developed by Janis Indrikis and Bob Cleave capitalizes on color-graphics displays and the pixel-manipulation capabilities of modern microcomputers. First, a Mollweide equal-area map projection of the entire surface of Earth is projected onto the computer screen. Then the coverage areas for the various satellite are superimposed onto that map projection. Finally, the computer makes binary pixel counts to determine and display the global coverage characteristics of that particular constellation.

THE SPACE EGGS SIMULATION PROGRAM

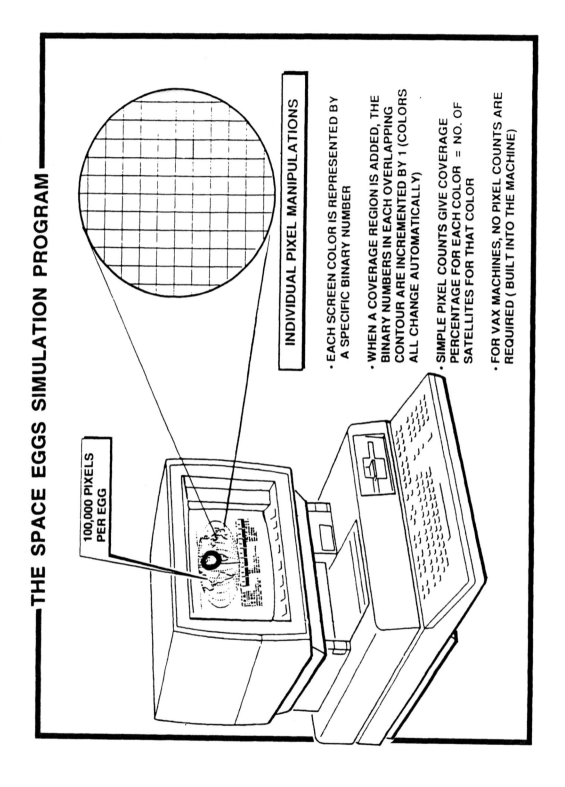

100,000 PIXELS PER EGG

INDIVIDUAL PIXEL MANIPULATIONS

- EACH SCREEN COLOR IS REPRESENTED BY A SPECIFIC BINARY NUMBER

- WHEN A COVERAGE REGION IS ADDED, THE BINARY NUMBERS IN EACH OVERLAPPING CONTOUR ARE INCREMENTED BY 1 (COLORS ALL CHANGE AUTOMATICALLY)

- SIMPLE PIXEL COUNTS GIVE COVERAGE PERCENTAGE FOR EACH COLOR = NO. OF SATELLITES FOR THAT COLOR

- FOR VAX MACHINES, NO PIXEL COUNTS ARE REQUIRED (BUILT INTO THE MACHINE)

219

In most cases, the footprints are simple horizon-to-horizon circles projected onto the spherical Earth, but the program can handle footprints with a variety of sizes and shapes. The coverage contours for a specific constellation consisting of 21 satellites in 550-nautical-mile orbits is shown at the bottom of Figure 9.6. The viewing contours are all circles, but they are distorted slightly when they are displayed on the Mollweide equal-area projection. The three different types of shading in the figure denote coverage regions with single, double, and triple satellite coverage.

The fine dot pattern highlights geographical regions with exactly one satellite in view. Those with more dense dot patterns have double satellite coverage, and those covering the black overlapping regions are within view of three satellites. The Space Eggs program is extremely flexible, it runs very rapidly, and it provides convenient human-oriented, color-graphics outputs.

The circular coverage footprints shown in Figure 9.6 are the most common coverage contours. But the Space Eggs simulation program can handle various other contours, including some that are shaped like wedges or flat donuts. When a simulation is complete, the program automatically produces summary displays pinpointing that fraction of Earth covered by different numbers of satellites over the interval being simulated.

CONSTELLATION SELECTION TRADES FOR MOBILE COMMUNICATION SATELLITES

Those of us who live on planet Earth are already being served by a number of satellite constellations, and, within a few short years, a variety of new ones will be launched into space to provide us with various other benefits including cellular telephone services for the entire globe.

Figure 9.6 In this "Space Eggs" computer simulation, circular coverage areas (footprints) are laid down on a Mollweide equal-area projection. For the case being simulated, 21 satellites are orbiting in a Walker constellation in 550-nautical-mile orbits with 27° inclinations. On the computer screen and in the hard-copy color-graphics outputs, the single-, double-, and triple-satellite coverage areas are all highlighted with their own unique colors.

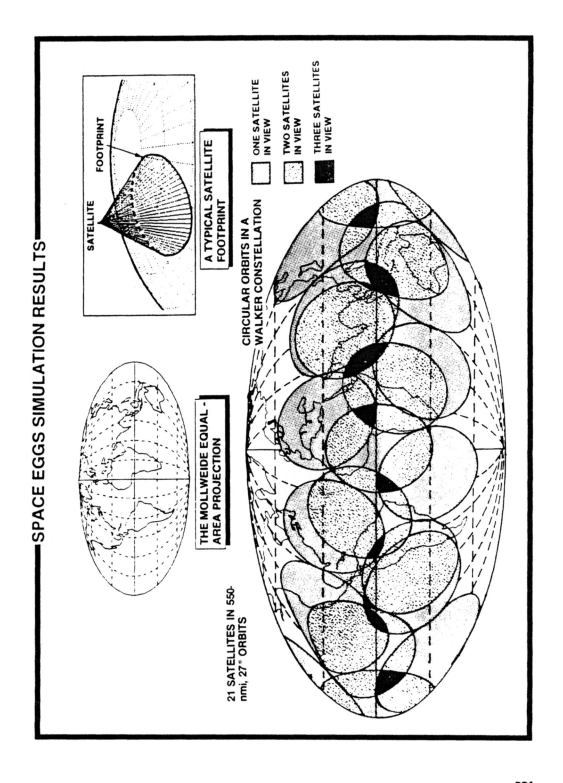

SPACE EGGS SIMULATION RESULTS

A TYPICAL SATELLITE FOOTPRINT

SATELLITE
FOOTPRINT

ONE SATELLITE IN VIEW

TWO SATELLITES IN VIEW

THREE SATELLITES IN VIEW

CIRCULAR ORBITS IN A WALKER CONSTELLATION

THE MOLLWEIDE EQUAL - AREA PROJECTION

21 SATELLITES IN 550-nmi, 27° ORBITS

These so-called mobile communication constellations are being targeted for three different orbital-altitude regimes:

1. Low-altitude orbits at approximately 400-nautical-mile altitude
2. Medium-altitude orbits in the 5000-nautical-mile range
3. Geosynchronous altitude constellations 19,300 nautical miles high

Choosing the best orbital-altitude regime for a proposed constellation of mobile communication satellites involves a number of subtle, but important engineering trades. Some of the critical selection criteria are highlighted in Table 9.2, which lists 12 different selection criteria—all of which are strongly dependent on the orbital-altitude regime chosen for the constellation. Notice how these 12 altitude-dependent characteristics have been grouped into three main categories centering around:

1. The orbital environment
2. The estimated cost of the constellation
3. Cost and complexity assessments for the overall system.

The rectangles with light dot patterns in Table 9.2 highlight the most desirable orbital-altitude regime in each row, whereas the darker dot patterns mark the worst flight regime associated with that particular characteristic. Unshaded regions are of intermediate desirability. These various selection criteria are further discussed in the next three sections, which also provide additional discussions of the characteristics of mobile communication satellite constellations.

The Orbital Environment

The low-level-altitude flight regime is the best of the three locations in terms of Van Allen radiation exposure and damage. This is true because the intrinsic shielding provided by Earth's magnetic field is the most effective at low altitudes. Mobile communication satellites positioned at the geosynchronous altitude also suffer rather minimal damage because they lie beyond the more intense portions of the upper Van Allen Radiation Belt. Those that are positioned in the intermediate-altitude flight regime, 5000 nautical or so miles above Earth, are in a slight notch between the two belts. Nevertheless, they suffer from the highest levels of radiation damage when compared with satellites in the other two flight regimes.

Periodic eclipse intervals cause the most severe problems for the low-altitude constellations, because low-altitude satellites enter Earth's shadow several times each day. On a typical mission, they spend at least 30 percent of their time in darkness. By contrast, the medium- and high-altitude satellites spend only about one or two percent of their time traveling through Earth's shadow.

Signal time delays, which can cause severe confusion and talk overlap, are, by far, the most severe for the high-altitude constellations because their electromagnetic signals must travel much longer distances. Low-altitude satellites generally experience the shortest signal-time delays.

Signal quality is degraded for small elevation angles because when a satellite is near the horizon, its signals are often blocked by natural and man-made barriers and they are distorted by Earth's atmosphere. Low-altitude satellites spend much of their time at low elevation angles and they sweep from horizon to horizon at a rapid rate. Medium-altitude satellites, by contrast, are usually situated well above the horizon and they move across the sky only about 1° per minute. A 1° arc corresponds to the thickness of a human finger held at arm's length. Geosynchronous satellites remain stationary in the sky, but users at high northern or southern latitudes tend to access them at low elevation angles down near the horizon.

Space debris fragments are, by far, the most numerous in the low-altitude flight regime. They occur in considerably smaller numbers at medium and geosynchronous altitudes. Hazardous collisions at these higher altitudes are thus much less frequent and they tend to be less energetic, too.

The Estimated Cost of the Constellation

The number of satellites required for continuous, global coverage is summarized in Figure 9.7 for the three orbital-altitude regimes under consideration. At the geosynchronous altitude, only five or six satellites are required for essentially continuous, global coverage. At the intermediate altitude 5000 nautical miles or so above Earth, 10 or 12 satellites are required to achieve the same level of coverage, whereas in the low-

TABLE 9.2 Constellation Selection Trades

■ Best Characteristics
■ Worst Characteristics

The Orbital Environment

Selection Criteria	Low-Altitude Constellations (~ 400 nmi)	Median-Altitude Constellations (~ 5,000 nmi)	High-Altitude Constellations (~19,300 nmi)
Van Allen radiation	Low levels of radiation in low-altitude orbits. Magnetic field shielding	Moderate levels of radiation in properly chosen orbits	Low levels of radiation in geosynchronous orbits
Eclipse intervals	Frequent day-night cycling. Satellite in darkness ~30% of time	Infrequent day-night cycling. Satellite in darkness ~ 2% of time	Infrequent day-night cycling. Satellite in darkness ~ 1 to 2% of time
Signal time delays	Shortest signal delay times ~ 0.02 sec for 2-way transmissions	Moderate signal delay times ~ 0.1 sec for 2-way transmissions	Longest signal delay times ~0.25 sec for 2-way transmissions
Spacecraft elevation angles	Rapidly varying elevation angles. Satellites frequently near horizon	Slowly varying elevation angles. Satellites well above horizon most of the time	No elevation angle variations. Satellites near the horizon for high-latitude users
Man-made space debris fragments	Large numbers of space debris fragments	Smallest numbers of space debris fragments	Moderate numbers of space debris fragments

Estimated Cost of the Constellation

Number of satellites required	Largest numbers of satellites required. Typically 30 to 60	Moderate numbers of satellites required. Typically 10 or 20	Smallest numbers of satellites required. Typically 3 to 6
Cost of each satellite and its transportation costs	Lowest-cost satellites: simple and light. Lowest cost transportation for each satellite	Moderate cost for each satellite. Moderate transportation cost for each satellite	Largest, most complex and costliest satellites. Highest transportation cost for each satellite
Projected satellite lifetime	Shortest satellite lifetime, typically 5 year life	Long satellite lifetime, typically 10 to 15 years	Long satellite lifetime, typically 10 to 15 years

Cost and Complexity Assessments for the Overall System

Are handoffs and/or crosslinks required?	Frequent handoffs and/or crosslinks required	Usually no handoffs or crosslinks required	No handoffs or crosslinks required
Cost and complexity estimates for the ground control segment	Usually the most complex and costly ground control links	Relatively low cost ground control segment	Relatively low cost ground control segment
Cost and weight of the personal communicators	Moderately costly, but lowest weight communicators	Moderately costly communicators of moderate weight	Inexpensive but heavy communicators
Can incremental startup coverage be achieved?	Incremental startup usually not practical	Incremental startup coverage can be practical	Incremental startup coverage very practical

THE APPROXIMATE NUMBER OF SATELLITES REQUIRED FOR CONTINUOUS, GLOBAL COVERAGE

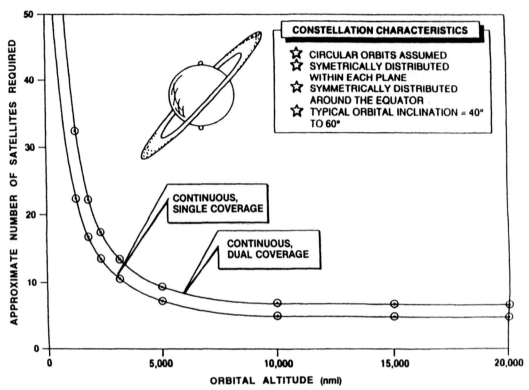

Figure 9.7 The approximate number of satellites necessary for single and double coverage at different altitudes above Earth are represented by these two smooth curves. In the high-altitude flight regime, only five or six satellites are required for essentially continuous global coverage. In the low-altitude flight regime, 400 to 500 nautical miles above Earth, 40 or more satellites are required to achieve roughly comparable coverage characteristics. At intermediate altitudes, 4000 to 5000 nautical miles high, 10 or 12 satellites in proper orbits will suffice.

altitude flight regime, 40 or more satellites are usually required.

Of course, the cost of building and launching each satellite is also a strong function of orbital altitude. The curves in Figure 9.8 provide us with estimates of the approximate cost of building each satellite and launching it to its desired orbital destination.

Low-altitude satellites are inexpensive to build and launch, but vast numbers of them are required to achieve continuous, worldwide coverage. Smaller numbers of high-altitude satellites are needed for comparable service, but each of them costs a huge amount of money to build and install. Consequently, as the curve on the right-hand side of Figure 9.8 indicates, the lowest overall cost can probably be achieved for constellations positioned within the intermediate flight regime.

As Table 9.2 indicates, the projected on-orbit satellite lifetime is generally the shortest for satellites at lower altitudes. At that altitude, a satellite typically lasts only five years or so. Their lives are shortened, in part, because they flash in and out of Earth's shadows so many times each day. These frequent day-night cycles place added burdens on a satellite's thermal control system and its batteries, which much undergo repeated charging and discharging cycles. Generally speaking, satellites tend to live longer in the intermediate and the geosynchronous flight regimes.

Cost and Complexity Assessments for the Overall System

The cost and complexity assessments for the overall system (see Table 9.2) is subdivided into four categories, the first of which centers around any requirements for hand-offs or cross-links. A hand-off is required when a satellite drops below the local horizon or provides the user with an unacceptably weak signal. When this happens, the communicator must switch to another satellite. Crosslinks are used to relay messages from one satellite to another in order to set up long-range telephone connections. Low-altitude satellites tend to require frequent hand-offs and/or crosslinks because each satellite covers such a small fraction of the globe. Medium- and high-altitude constellations rarely require hand-offs.

As Table 9.2 indicates, the cost and complexity of the ground control segment tends to be the highest for the low-altitude constellations because their satellites are more numerous and because they travel across the sky at such a rapid rate.

The weight of the personal communicators is generally the lowest for the low-altitude flight regime, but such communicators tend to be moderately costly. Those that communicate with satellites in the medium- and high-altitude flight regimes are usually heavier and bulkier.

ROUGH ORDER-OF-MAGNITUDE COSTS FOR BUILDING AND INSTALLING MOBILE COMMUNICATION CONSTELLATIONS AT VARIOUS ALTITUDES

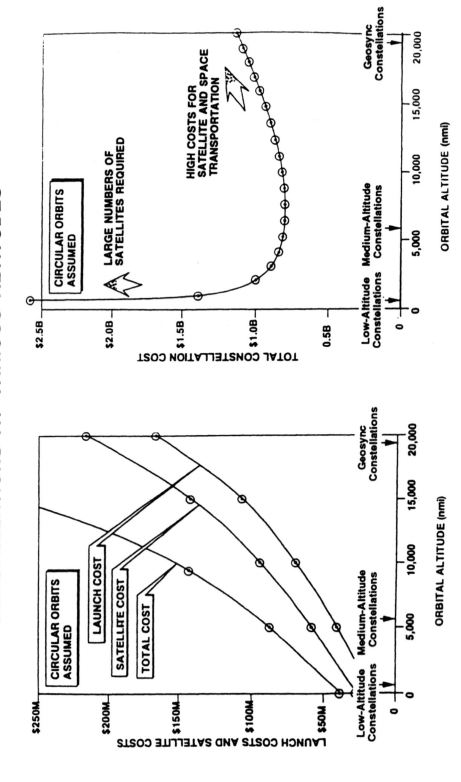

Those individuals who finance mobile communication constellations, in some cases, can launch a few satellites into orbit and derive income from their use while they are installing the rest of the constellation. If this is possible, their business is much more likely to succeed. Low-altitude constellations usually have little potential for bringing in money from incremental startup coverage. Those launched into medium- and high-altitude orbits have a much better potential for collecting incremental startup fees. This potential is, of course, the greatest for constellations at the geostationary altitude because when only one geostationary satellite has been launched, it can provide useful incremental coverage continuously for a specific portion of the globe.

THE GLOBAL BENEFITS OF MOBILE COMMUNICATION SATELLITE CONSTELLATIONS

In the early days of the space program, aerospace engineers were happy if they could hurl a satellite into space that did not fall immediately back to Earth. A few years later, we were routinely launching our satellites into specific orbits with extremely precise orbital altitudes and inclinations. More recently, we have learned how to launch swarms of satellites that fly in precise, evenly spaced constellations. Some of these constellations are already providing useful services for people scattered around the globe.

Within a few years, widely dispersed mobile communication satellite constellations will allow users on the ground to make telephone calls from anywhere to anywhere. When this exciting technology is in service, a researcher or an adventurer

---

Figure 9.8 As these parametric curves indicate, a constellation of mobile communication satellites launched into high-altitude geosynchronous orbits tends to cost more than a comparable constellation occupying the medium altitude flight regime because of the high cost of building and launching each geosynchronous satellite. In the low-altitude flight regime, each satellite is cheaper to build and launch, but extremely large numbers of them are required for continuous global coverage. In the intermediate-altitude regime, 5000 to 6000 nautical miles high, the lowest cost compromise constellation probably can be achieved.

lollygagging on the Serengetti Plain will be able to chat with friends and colleagues at an Antarctic research station or an unpopulated Pacific island. Interconnections of this type are just now becoming practical because of today's emerging mobile communication satellite constellations.

10 SPACE-AGE TECHNOLOGIES FOR THE 21ST CENTURY

The Newtonian principle of gravitation is now more firmly
established on the basis of reason than it would be were the
government to step in, and make it an article of necessary faith.
Reason and experiment have been indulged,
and error has fled before them.
—*Thomas Jefferson* (Notes on the State of Virginia, *1781-1785*)

In working your way through the first nine chapters of this
book, you may have picked up the erroneous impression that
there is not much left to do in the field of orbital mechanics.
Actually, a number of orbital mechanics concepts are awaiting
effective exploitation. Consequently, at this juncture, it seems
mandatory to discuss some of these advanced concepts to help
inspire and motivate a new generation of ambitious entrepreneurs and aerospace engineers. Enormous fortunes definitely will be made along the space frontier. For, as space
industrialization expert Art Dula has observed, "The next
generation of billionaires is going to come from the business
of space."

This final chapter will introduce you to a sampling of these
cutting-edge technologies, including chemical and electro-
chemical mass drivers, nuclear- and laser-powered rockets,
solar sailing techniques, tethered satellites, and flexible skyhook cables draped down toward Earth from the geosynchronous altitude. Any one of these rather esoteric concepts,
properly exploited, could provide the next generation of aerospace professionals with ample intellectual stimulation to
keep them working and dreaming well into the twenty-first
century—and beyond.

CHEMICAL MASS DRIVERS

Shortly after the end of the Civil War, science fiction writer Jules Verne published two short novels dealing with the construction of a gigantic cannon specifically designed to hurl small teams of astronauts on round-trip journeys around the Moon. Jules Verne was a Frenchman who never ventured across the Atlantic Ocean. But he prophetically concluded that the Americans, with their abiding enthusiasm for cutting-edge technology, would end up visiting our nearest celestial neighbor in space.

The launch site Verne selected was in southern Florida not far from present-day Cape Canaveral. At that unlikely location, his American protagonists started their grand adventure by digging a 900-foot tunnel straight down toward the center of Earth. Then they filled the first 200 feet with a superpowerful gunpowder specially formulated for their upcoming Moon mission. Three courageous astronauts then climbed into a chubby, bullet-shaped capsule that has lowered down the tunnel and then fired toward the Moon by the force of the explosion. Two plump chickens and a mongrel dog accompanied Verne's hapless astronauts who were dressed in the typical business suits of the day.

Jules Verne knew that extremely high velocities would be required to hurl his three brave astronauts to the Moon, but he paid hardly any attention to the deadly g forces associated with their abrupt departure. His descriptions of weightlessness are surprisingly authentic. But he incorrectly concluded that the astronauts would be weightless only during a brief instant, when they passed through the point at which the gravitational fields of Earth and the Moon were pulling on their bodies with equal forces of attraction.

The characters in this story were also largely oblivious to the many of the subtleties of systems engineering. Minutes before blastoff, his three hapless astronauts are still engaging in a protracted argument about how they should position their bodies inside their padded capsule to best resist the g loads they know they will soon encounter.

As the projectile is headed toward the Moon, nearby astronomers are getting ready to observe it through their telescopes. Unfortunately, the black cloud of powder spewed out by the explosion coats the lenses of their telescopes so they are not able to see anything.

Modern space-age rockets, with their short burning arcs, closely approximate the explosive acceleration of the Jules Verne cannon. When we launched the Apollo astronauts toward the Moon, the burning interval of their Saturn V rocket totaled only about 15 minutes. This was followed by a three-day interval during which they coasted through cis-lunar space.

In the 1960s the Canadian munitions specialist, Gerald Bull, conducted a series of experiments in which he attempted to demonstrate that a modified 16-inch naval gun could be rigged to hurl small rockets onto space-bound trajectories. According to his careful calculations, his projectiles would be subjected to about 2000 g's, so only the most rugged payloads would survive. When the rocket breached the atmosphere, it was to be ignited to provide the final velocity increment. Gerald Bull's engineering calculations indicated that if his cannons were fired often enough, they could place 100-pound payloads into orbit for only about $500 per pound. This is roughly five or six times less costly than the most efficient payload delivery of today.

ELECTROMAGNETIC CATAPULTS

Someday electromagnetic catapults may be used in hurling small payloads into space. Several years ago such a concept was courageously proposed by Gerard O'Neill of Princeton University. O'Neill's main goal was to install a large electromagnetic catapult on the surface of the Moon. Once it was in place, it would hurl 40-pound masses up to a velocity of about 1.5 miles per second, the surface escape speed of the Moon. O'Neill estimated that a 20,000-ton mass driver on the Moon, over a period of several years, could, hurl 1000 times its mass up to escape velocity at a cost of only about 50 cents per pound.

A large, soft "catcher's mitt" positioned above the backside of the Moon at the L_2 libration point would capture the payloads in space. Later, they were to be towed to the geosynchronous attitude where they would be used in constructing gigantic space colonies.

Larger electromagnetic catapults have been considered for use in launching payloads from the surface of Earth. In one proposal, an electromagnetic coil gun costing $1.3 billion would accelerate 1000-pound projectiles to a velocity of 19,000 feet per second. Each projectile would consist of a rugged 200-

pound payload attached to an 800-pound rocket that would add enough velocity to deliver the payload into the desired orbit. Peak accelerations generated by the 2500-foot coil gun would amount to approximately 1000 g's.

NUCLEAR PROPULSION

When we use a chemical rocket to deliver a payload into space, the amount of energy we can impart is intrinsically limited by the amount of energy contained in its chemical propellants. One way to circumvent this difficulty is to use nuclear energy instead.

Project Orion

In the 1960s, Freeman Dyson proposed a novel method for using nuclear energy in propelling payloads through space. In Dyson's proposal, which was called Project Orion, large space vehicles would be accelerated with a series of spaceborne nuclear explosions. As Figure 10.1 indicates, each nuclear device is detonated behind a rigid pusher plate connected to the space vehicle with big, heavy springs specifically designed to absorb the abrupt shock of the explosions.

Between each pair of explosions, the pusher plate is coated with a thin film of oil. This helps protect the pusher plate from the blast and heat generated by the next explosion. Structural damage is further reduced by purposely employing very low-yield nuclear explosions with the energy equivalent of only 10,000 *pounds* of TNT.

The Nerva Rockets

Rockets designed to harness the thermal energy from controlled nuclear reactions have also been built and test-fired on static test stands. Specifically, in the 1970s, the nuclear-powered Nerva rocket was tested in a series of static firings at Jackass Flats, Nevada. It generated 75,000 pounds of thrust at an average specific impulse of about 700 seconds. Unlike chemical rockets, the efficiency of nuclear-powered rockets is not limited by the energy their propellants contain. Consequently, it can be considerably more efficient (higher specific impulse).

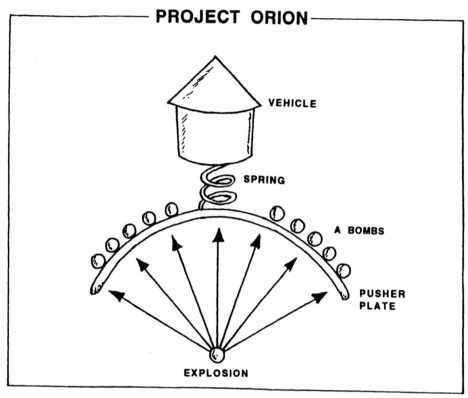

PROJECT ORION

VEHICLE

SPRING

A BOMBS

PUSHER
PLATE

EXPLOSION

Figure 10.1 *The well-known physicist Freeman Dyson proposed a simple nuclear-powered rocket to be propelled through interstellar space by a series of nuclear explosions detonated behind a thick, metal pusher plate. Heating of the pusher plate would be mitigated by purposely employing low-yield nuclear explosions and by coating the metal plate with a thin film of oil immediately before each new explosion.*

For many years, nuclear rocket technology has been quiescent, but new design concepts have been proposed for possible use in connection with manned, interplanetary space probes bound for Mars. According to a recent article in *Aerospace America*, a nuclear thermal rocket of modern design could provide specific impulse values ranging between 800 and 1000 seconds. Such a rocket could carry about twice as much payload onto a trans-Mars trajectory compared with a chemical rocket weighing the same and of a similarly sophisticated design.

LASER-POWERED ROCKETS

A number of designs for laser-powered rockets (see Figure 10.2) have been proposed for propelling payloads into space. In such a scheme, a ground-based laser is rigged to beam electromagnetic energy to the base of a rocket where it vaporizes a hydrogen-rich compound to create thrust.

A few small-scale studies have been conducted to determine the feasibility and the practicality of laser-powered rockets. In one study, Avco engineer Dennis Riley proposed that pairs of closely spaced laser pulses be employed. The first pulse would vaporize the material; the second, more powerful pulse would cause the vapor to expand, thus creating the desired thrust.

Riley's study team concluded that a well-designed laser-powered rocket might be able to deliver 15 percent of its liftoff weight into a low-altitude Earth orbit. According to their engineering estimates, a small-scale facility kept in continuous operation would be able to orbit small payloads for as little as $45 per pound.

SOLAR SAILS

Solar sails assembled in various configurations have been proposed for raising the energy of orbiting payloads so they could escape Earth or fly into higher-altitude orbits. As Figure 10.3 indicates, a solar sail is a thin filmlike sheet that uses solar-radiation pressure as its means of locomotion.

In the late 1980s, several research centers were planning to enter a "solar sailing competition" to commemorate the 500th anniversary of Columbus's journey to the North American continent. After executing a precise swing-by maneuver around the Moon, their destination was to be the red planet, Mars. Preliminary estimates indicated that the proposed solar sails could be constructed for $3 million to $15 million each.

The propulsive force obtained from the Sun at a distance of one astronomical unit amounts to only about five pounds per square mile. However, despite its seemingly small magnitude, such a tiny force can add substantial amounts of energy to deep-space payloads because they are accelerated so relentlessly for such a long period of time.

LASER-POWERED ROCKETS

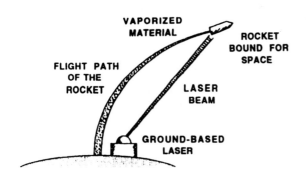

VAPORIZED MATERIAL

ROCKET BOUND FOR SPACE

FLIGHT PATH OF THE ROCKET

LASER BEAM

GROUND-BASED LASER

PULSING THE ROCKETS

☆ AVCO ENGINEER DENNIS REILLY HAS PROPOSED:

 • TWO CLOSELY SPACED PAIRS OF PULSES

 • 1st PULSE VAPORIZES THE MATERIAL

 • SECOND, MORE POWERFUL, PULSE CAUSES THE VAPOR FILM TO EXPAND

☆ COMPUTER SIMULATIONS INDICATE THAT:

 • LASER-POWERED ROCKET MAY BE ABLE TO DELIVER 15% OF ITS LIFTOFF WEIGHT INTO ORBIT

LASER-POWERED ROCKET'S THEORETICAL CAPABILITIES

☆ RESTRICTED TO SMALL, COMPACT PAYLOADS

☆ CONTINUOUS OPERATION = 64,000 POUNDS PER YEAR IN ORBIT

☆ AMORTIZED R&D PLUS OPERATING COSTS YIELDS THEORETICAL $45 PER POUND ORBITED

Figure 10.2 In theory, powerful ground-based lasers could be used to propel small payloads from the launch pad upward into outer space. As this conceptual diagram indicates, continuous propulsion can be achieved by using the ground-based laser to vaporize hydrogen-rich compounds attached to the rear of the rocket.

The Soviet Regatta Satellites

Russia's aerospace engineers have successfully used solar-radiation pressure to provide gentle corrective forces to control the orientations of some of their orbiting satellites. The necessary sensors, hardware, and software have been installed aboard their so-called Regatta satellites.

Small, flat vanes positioned around the periphery of each Regatta satellite swivel and tilt to catch the Sun's rays and, in turn, control the vehicle's attitude in real time. Russian entrepreneurs are attempting to license this technology through Princeton University's Space Studies Institute.

A SAILING RACE TO MARS

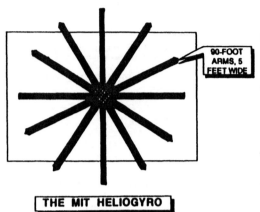

90-FOOT ARMS, 5 FEET WIDE

THE MIT HELIOGYRO

COST ESTIMATES

☆ SOLAR SAIL COST RANGES FROM $3M TO $15M

☆ $10M NEEDED TO SET UP THE COMPETITION

COLUMBUS 500 SPACE SAIL CUP

☆ COMPETITION TO COMMEMORATE 500th ANNIVERSARY OF THE CHRISTOPHER COLUMBUS VOYAGE

☆ SUNLIGHT TO PROPEL THE CONTESTANTS PAST THE MOON TOWARD MARS

THE VARIOUS VEHICLES

☆ WORLD SPACE FOUNDATION, PASADENA, CALIFORNIA

☆ JOHNS HOPKINS UNIVERSITY APPLIED PHYSICS LABORATORY

  • 560-FOOT SAIL RIMMED WITH 4-FOOT "PETALS"

☆ MASSACHUSETTS INSTITUTE OF TECHNOLOGY

  • "HELIOGYRO" WITH 90-FOOT ARMS

Figure 10.3 A thin plastic sheet coated with a single molecular layer of aluminum could become a solar sail capable of propelling an inexpensive space probe onto an escape trajectory or toward distant planets such as Jupiter or Mars. A competition intended to commemorate the 500th anniversary of the first "New World" voyage of Christopher Columbus was planned for 1992. It drew interest from three potential contestants from new-world organizations but, unfortunately, none of the contestants was able to raise enough money to construct the hardware needed to complete the mission.

Figure 10.4 In theory, entirely new types of ultrathin "geostationary" satellites could be positioned in "orbits" in which they would experience exactly the proper combination of gravity and solar-radiation pressure to keep them at their assigned locations. Such a "satellite" must be placed in an "orbit" that is inclined at least 23.5° with respect to Earth's equator. Otherwise, it will dip periodically into Earth's shadow thus nullifying the counterbalancing solar-radiation pressure.

"GEOSTATIONARY" SATELLITE VIA SOLAR-RADIATION PRESSURE

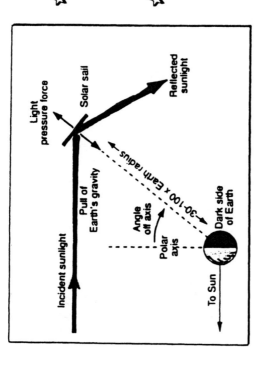

THE POLESAT GEOMETRY

☆ TYPICAL "INCLINATION" ANGLES = 30° TO 40°

- ANGLE MUST EXCEED 23.5°

- 90° ANGLE IS POSSIBLE (POLESAT HOVERS OVER THE POLE)

☆ TYPICAL ALTITUDES (AREA/WEIGHT-DEPENDENT) = 30 TO 100 EARTH RADII

- SIGNAL TRAVEL TIMES (TWO WAY) = 1.3 TO 4.2 SECONDS

- OVER-THE-POLE POLESAT = 270 EARTH RADII (TWO-WAY SIGNAL TRAVEL TIME = 11.5 SECONDS)

THE POLESAT CONCEPT

☆ EARTH'S GRAVITY COUNTERBALANCED BY SOLAR-RADIATION PRESURE

☆ DEVELOPED BY ROBERT FORWARD AT HUGHES RESEARCH LABS (PATENT PENDING)

"Geostationary Polesats"

The pressure of the Sun shining on a flat surface also can be used to create a new kind of "geostationary" satellite. Figure 10.4 shows how such a stationary solar sail can hover at about 30 to 100 radii from Earth above any latitude exceeding 23.5°.

Notice how the solar sail is tipped at an oblique angle. This creates a delicate balance between the gravitational force pulling the solar sail toward the center of Earth and the solar-radiation pressure pushing it in the opposite direction.

Robert Forward and his colleagues at Hughes Research Laboratory in Malibu, California, have applied for a patent on this so-called "Polesat" concept. Polesats could be used for a number of practical applications, but real-time communications relay is probably not among them. A practical Polesat must be so far away from Earth it would require 1.3 to 4.2 seconds for a modulated electromagnetic wave to make the necessary round-trip journey.

If they are boosted into higher-altitude locations, Polesats can be positioned directly over the North Pole or the South Pole. However, according to the engineering calculations put together by Robert Forward, the minimum altitude for such an installation is about 270 Earth radii, or about 1 million miles above Earth.

TETHERED SATELLITES

For many years tethered satellites have been of theoretical interest to aerospace engineers. However, only recently have tethered-satellite deployments been attempted. Two large tethers were carried into orbit aboard the shuttle orbiter; a few smaller ones have gone into space on expendable Delta boosters.

Some of the characteristics of a typical tethered satellite are summarized in Figure 10.5. Two versions were produced by

---

Figure 10.5 Long tethers draped down from the space shuttle can be used to generate electricity, scoop up tenuous atmospheric samples, and perform surprisingly efficient on-orbit maneuvers. Ambitious experimental tethered satellites were carried aloft on two separate space shuttle missions. On the first mission, the plastic tether snagged when it had been reeled out about 800 feet. During the second mission, the tether snapped in two when 11 miles of it had been successfully deployed.

TETHERED SATELLITES

PRACTICAL BENEFITS OF TETHERED SATELLITES

☆ GENERATING ELECTRICAL POWER

☆ DEORBITING OR REBOOSTING PAYLOADS

☆ STUDYING EARTH'S UPPER ATMOSPHERE

☆ PRECISE MICROGRAVITY CONTROL

TETHER CHARACTERISTICS

☆ 0.1-INCH DIAMETER COPPER, TEFLON, KEVLAR, AND NOMEX TETHER

☆ WEIGHS 29 POUNDS PER MILE
TENSILE STRENGTH = 400 POUNDS
(FULLY DEPLOYED TETHER = 13-POUND FORCE)

THE FIRST TETHER TEST

☆ SHUTTLE ATLANTIS 12-MILE-LONG ELECTRICALLY CONDUCTIVE CABLE 0.1 INCH IN DIAMETER

☆ HALF-TON SPHERICAL TETHERED SATELLITE 5000 VOLTS DOWNWARD

☆ ITALY'S ALENIA ESPAZIO SPENT $191M ON THE LAUNCH

☆ A COMPLETE PENDULUM CYCLE = 45 MINUTES

"TETHERS CAN EVEN BE USED AS VIRTUAL SKYHOOKS TO LOFT SATELLITES FROM EARTH'S SURFACE AND SWING THEM IN ARCS TO GEOSYNCHRONOUS ORBITS"

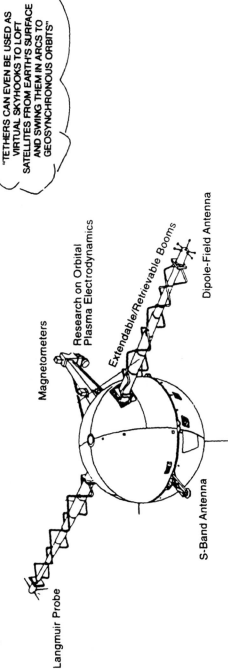

Magnetometers

Research on Orbital Plasma Electrodynamics

Extendable/Retrievable Booms

Dipole-Field Antenna

Langmuir Probe

S-Band Antenna

Italy's Alena Espazio with the assistance of various international partners. The Teflon/mylar tether they constructed was 12 miles long and about one-tenth of an inch in diameter. That electrically conductive tether was designed to hold a half-ton satellite vertically above the space shuttle. Unfortunately, the first tether had been reeled out only about 800 feet before it snagged. In the second attempt, the tether reached a range of more than 11 miles before it snapped in two.

The tensile strength of the two tethers was about 400 pounds, but even when they were fully deployed, a force of only about 13 pounds was expected to tug on the tether. The dynamic real-time forces acting on a tethered satellite are enormously complicated. Consequently, the engineers who worked on the tether struggled to predict the dynamics of the satellite and its tether during deployment. At this writing, the most likely explanation for the failure of the second tethered satellite is an interior flaw in the tether which caused electrical sparking to burn its outer sheath in two.

PROJECT SKYHOOK

In the early 1960s when Project Apollo was in full flower, my friend Bob Africano and I encountered a small announcement in the company newspaper at North American Aviation. According to the information it contained, our Public Relations Department was recruiting aerospace engineers to teach space-related courses to a small group of high school students at the California Museum of Science and Industry near downtown Los Angeles.

The Fundamentals of Space Exploration

With great enthusiasm, Africano and I began to outline a course on aerospace technology we were hoping to teach. And we became even more enthusiastic when we stumbled on the perfect title: "The Fundamentals of Space Exploration."

Every Tuesday evening at the museum we were surrounded by a ragged cluster of high school students—an equal mixture of girls and boys. Of course, they had not yet studied calculus or differential equations, so we were constantly striving to find interesting ways to illustrate the basic concepts of our craft using largely nonmathematical approaches.

In one early session, for instance, I proposed a novel technique for launching a satellite in which we would drop it into orbit from the top of a rigid 22,300-mile-high tower. Our students were fascinated by the fact that such a satellite would not fall toward the ground. Instead, it would hang motionless just off the ends of our fingertips. We also explained to them that if they dropped a satellite from the tower at any altitude above 16,100 miles, it would automatically swing into an elliptical orbit.

The Indian Rope Trick

One morning, Bob Africano ambled into the office with a big grin on his face. "The building would crush the bedrock," he said. "How can we solve that problem?"

"Why not build the tower higher," I replied. I then went on to explain that if the construction crew would extend the tower above the normal 22,300-mile altitude, the centrifugal force pulling up on its new materials above geosynchronous altitude would exceed the gravitational force pulling them downward! Consequently, the taller we would build the tower, the lighter it would be. Within a few minutes, we had calculated that a tower of uniform cross section 100,000 miles high would not weigh anything at all!

One morning, Bob arrived at work with a John Wayne swagger. "If we tear down the tower," he told me, "its elevator cable would hang suspended upward in space."

At first this seemed totally unbelievable to me. "I don't understand what you're driving at," I told him. "What would keep it in place?"

Africano then went on to explain that everywhere—all along its length—the entire skyhook cable is always in tension. No crosswise or compressive forces exist anywhere along it, so it doesn't need to be rigid to remain in place. We were both profoundly intrigued by this curious concept as new things to figure out danced through our brains. "The Indian rope trick could have been done all along," we later concluded. "It's just that everyone was using ropes that were too short."

Tapering the Cable

Unfortunately, when we calculated the mechanical stresses that would exist at various points along the cable (see Figure 10.6), our excitement damped down a little. To our dismay we

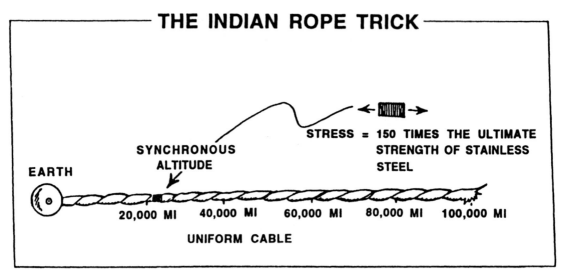

Figure 10.6 If we can get this uniform, 100,000-mile cable into position vertically above the equator, it will hang motionless in space with its lower end stretching down to the ground. Once it has been anchored in place, we can use the skyhook to hoist payloads into medium-altitude elliptical orbits or to fling them toward the distant corners of the universe.

learned that at the maximum stress on the cable, which would occur at the geosynchronous altitude, would be 150 times the ultimate strength of stainless steel. Thus, we concluded that even if we could somehow get the cable in place, it would immediately snap in two at that critical location.

However, even if it couldn't be built with existing materials, the skyhook was a terrific teaching tool, so we incorporated these concepts, and various others, into our lectures in that dilapidated basement classroom at the California Museum of Science and Industry.

Several months later, one of the company engineers brought us a newspaper article describing the research efforts of John D. Isaacs at Scripps Institute of Oceanography in La Jolla, California. Isaacs, it turned out, had independently proposed the installation of a flexible skyhook cable. Moreover, he had, to our surprise, also solved the seemingly intractable problem associated with the excessive stresses in the cable.

In his article, which was published in *Nature* magazine, Isaacs pointed out that the cable could be *tapered*, thus making it stronger at the point of maximum stress (see Figure 10.7). He also showed that, if it was constructed from whisker-thin single

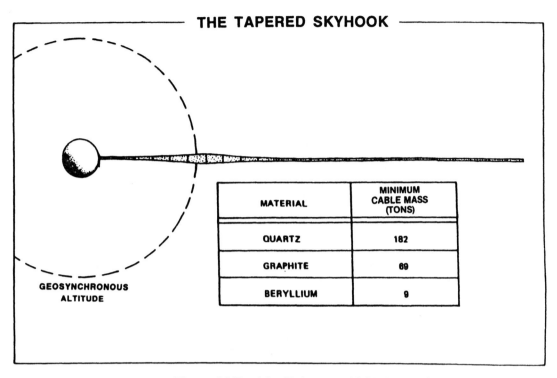

THE TAPERED SKYHOOK

MATERIAL	MINIMUM CABLE MASS (TONS)
QUARTZ	182
GRAPHITE	69
BERYLLIUM	9

GEOSYNCHRONOUS ALTITUDE

Figure 10.7 John D. Isaacs, a highly innovative and respected professor at Scripps Institute of Oceanography, proved mathematically that a properly tapered skyhook cable a bit longer than 23,000 miles could support its own weight without breaking. Theoretically, skyhook cables constructed from pure whiskerlike crystals of quartz, graphite, or beryllium would weigh less than 200 tons each.

crystals of quartz, graphite, or beryllium, the entire cable might weigh 200 tons or less.

Later, at the Lake Arrowhead Conference Center operated by UCLA, we met John Isaacs at a weekend short course. He was a gregarious individual with piercing blue eyes and a big, white, bushy beard. When Africano asked him how he had managed to come up with the idea of tapering the cable, he indicated that he had no difficulty with that idea because "marine cables are frequently tapered for exactly the same reason."

When we informed him that we had developed the idea for the flexible skyhook 18 months before his initial concepts, he merely shrugged. Russian researchers, it turned out, had developed essentially the same concepts six years earlier than we had in the days when Africano and I were still undergraduates struggling to understand the rudiments of differential calculus.

Dropping Satellites Into Orbit

But no matter who had figured it out first, our students loved to listen to our enticing lectures dealing with the wonders of Project Skyhook. Their eyes sparkled with enthusiasm as we told them about a young astronaut who was assigned to carry a satellite from the surface of Earth up to the top of the cable. On the ground, of course, he would experience his normal body weight. But as he climbed higher and higher up along the cable, his body would get lighter and lighter until he arrived at the geosynchronous altitude—where his body and the satellite he was carrying would become weightless.

As he continued to climb upward above that critical altitude, his head would point toward the center of Earth and he would have to reverse his movements and climb "downward" to go "up" the stairs! This curious reversal occurs because at any point along the cable above the geosynchronous altitude, the centrifugal force hurling his body outward exceeds the gravitational force pulling him down toward the center of Earth.

As we pointed out to our students, this variation in the effective "gravity" along the skyhook would allow us to construct realistic planetary museums at various altitude locations. The gravitational force at the surface of Mercury, for example, is 0.365 g. Consequently, we could build a realistic Mercury museum on the skyhook about 5000 miles above the surface of Earth. Similarly, Mars has a surface gravity of 0.3787 g, so its planetary museum would be located on the skyhook at the 3000-mile altitude.

THE SKYHOOK PIPELINE

Our students, who outfitted themselves in paisley blouses and thick leather belts, loved to listen to Sonny and Cher on their plastic portable radios. They also enjoyed hearing our stories about the skyhook complex and the possibility that someday, it might become our fifty-first state.

Once future engineers have installed a sufficiently strong skyhook cable and anchored it to the bedrock, they can elongate it to an altitude of 100,000 miles or more, hoist satellites along its length, and drop them into various elliptical orbits. At today's electrical rates, the energy required to hoist a one-pound satellite up to the minimum drop altitude to achieve orbit (16,100 miles) costs only a few pennies per pound. Optical laser stations positioned on the ground will beam energy to special elevators equipped with linear induction motors that pull them up along the sides of the cable.

Satellites will not be the only useful payloads hoisted up the skyhook aboard the laser-powered elevators. The elevators will be used to hoist radioactive waste products up to an altitude of 32,000 miles. At that altitude, or beyond, any objects we release will reach escape velocity and swing into harmless orbits around the Sun.

The laser-powered elevators also will be used to carry new structural materials to beef up the skyhook cable so it can support heavier loads. Then, once it is sufficiently strong, the elevators will carry hollow pipeline sections vertically upward along its length until six long, skinny pipelines have been rimmed around its periphery (see Figure 10.8). These six parallel pipelines will stretch from the surface of Earth to an altitude of 130,000 miles.

Seawater then will be pumped up through the pipelines to the geosynchronous altitude. Once it passes the 22,300-mile altitude (see Figure 10.9), the water will "fall" vertically upward to spin electrical turbines. By the time the water reaches an altitude of 96,000 miles, it will yield just enough energy to pay off the pumps. Beyond that altitude, the pipeline produces net useful electrical energy. When each pound of seawater reaches an altitude of 130,000 miles, it will have contributed the net energy equivalent of a pound of gasoline. Thus, in effect, the Skyhook Pipeline will allow us to convert all the water in our oceans into gasoline! This pure, pollution-free electricity can be beamed from the turbine sites back to Earth to power the wheels of our civilization.

Some of our students wondered what would happen as the pipeline carried increasingly larger amounts of water out of the oceans. Over time, the level of the oceans would drop slightly. Surprisingly, however, a single inch of ocean water would provide all of the world's global energy needs, at present usage rates, for more than 2000 years. The average depth of the oceans is about two miles. So, even if we produced all of our energy using the Pipeline, we would have enough seawater to last hundreds of thousands of years.

"But where does the energy come from?" they would sometimes ask. "Isn't this a new kind of perpetual motion machine?" Not at all. Earth is a giant flywheel. It weighs 13 million billion billion (13×10^{24}) pounds. Locked within its spinning mass is enough energy to power the wheels of our civilization for 860 million years.

THE CONSTRUCTION OF THE SKYHOOK PIPELINE

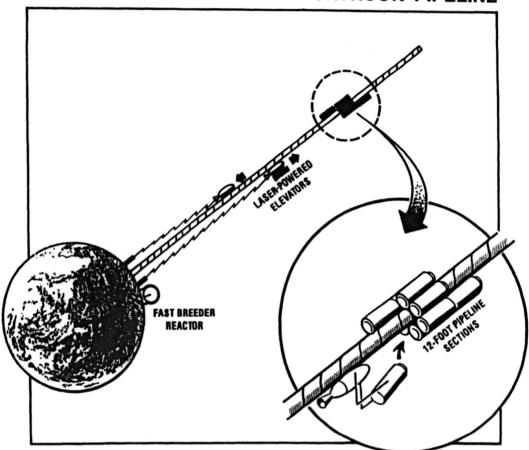

Figure 10.8 Once a sufficiently strong skyhook cable has been constructed straight up above the equator, dedicated, laser-propelled elevators can carry additional cabling materials upward along its length to make it both longer and stronger. Cylindrical pipeline sections then can be hoisted in place to provide six parallel pipelines capable of pumping vast quantities of water from the oceans into space.

THE SKYHOOK COMPLEX

Throughout the twenty-first century, as the Skyhook Pipeline expands its operations, we can begin expanding it into the Skyhook Complex, one version of which is shown in Figure 10.10. Notice that a number of profit-making facilities have been attached to the pipeline mostly at the geosynchronous altitude. At that altitude, we can attach as much extra mass as we like without increasing the stresses on the cable. These new

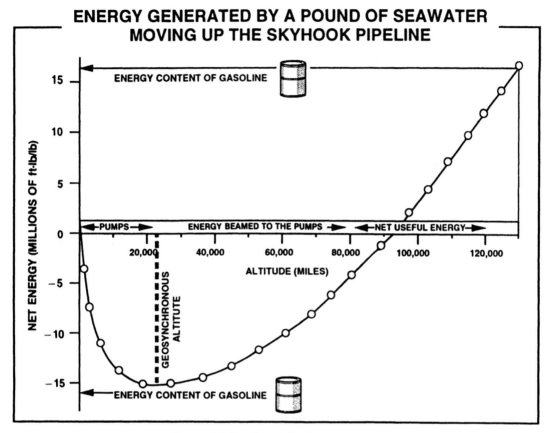

ENERGY GENERATED BY A POUND OF SEAWATER MOVING UP THE SKYHOOK PIPELINE

Figure 10.9 Pumping a pound of seawater from the surface of Earth up to the geosynchronous altitude requires a total energy expenditure of approximately 16 million foot-pounds. Above the geosynchronous altitude, the centrifugal force exceeds the pull of gravity, so the water falls *vertically upward* from that point on. By the time it reaches an altitude of 130,000 miles, each pound of seawater can be converted into the net energy equivalent of a pound of gasoline.

facilities will include hospitals for treating burn victims and paraplegics. Those who are suffering from chronic ulcers, heart disease, hypertension, joint problems, and strokes also can be treated beneficially at the skyhook medical clinics.

Tourist hotels and large-scale factories will also be attached to the pipeline at the geosynchronous altitude, and various other facilities will be installed at other locations. These will include a self-propelled flight dome at the 5000-mile altitude where the net *g* loads amount to only 10 percent of normal gravity. At that altitude, even the flabbiest stock broker would be able to don a pair of artificial wings and fly around in graceful

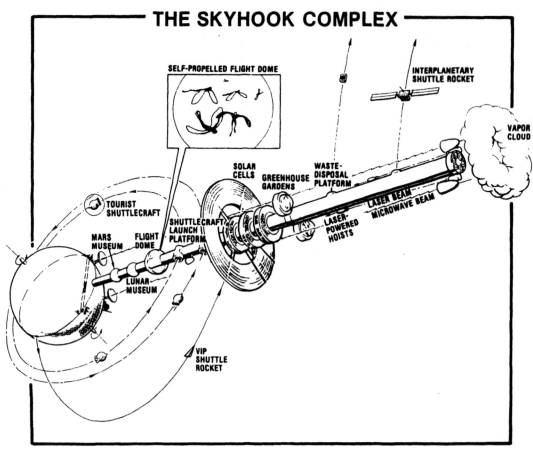

THE SKYHOOK COMPLEX

SELF-PROPELLED FLIGHT DOME

INTERPLANETARY
SHUTTLE ROCKET

VAPOR
CLOUD

SOLAR
CELLS

GREENHOUSE
GARDENS

WASTE-
DISPOSAL
PLATFORM

LASER BEAM

MICROWAVE BEAM

TOURIST
SHUTTLECRAFT

SHUTTLECRAFT
LAUNCH
PLATFORM

LASER-
POWERED
HOISTS

MARS
MUSEUM

FLIGHT
DOME

LUNAR-
MUSEUM

VIP
SHUTTLE
ROCKET

Figure 10.10 By the time the Skyhook Complex has been completely installed in space, it could become the fifty-first state. Large factories, hotels, hospitals, and greenhouse gardens can be attached to the Pipeline at the geosynchronous altitude without adding any additional stress. By the dawning of the twenty-second century, earthlings will probably wonder how their ancestors could have managed to exist without the goods and services provided by the Skyhook Complex.

Figure 10.11 Tourists who are fortunate enough to visit tomorrow's Skyhook Complex will be able to board a shuttlecraft for a quick trip around Earth. During the 24-hour journey, their unpowered shuttlecraft will be traveling along a totally predictable trajectory while those onboard will be drinking vintage wines and eating gourmet foods. They will also enjoy wide-screen, multichannel, laser video disks carefully synchronized with the interesting tourist spots they are sailing over. One day later, at apogee, their shuttlecraft can be gently captured at precisely the same altitude at which it was released in a big, soft net.

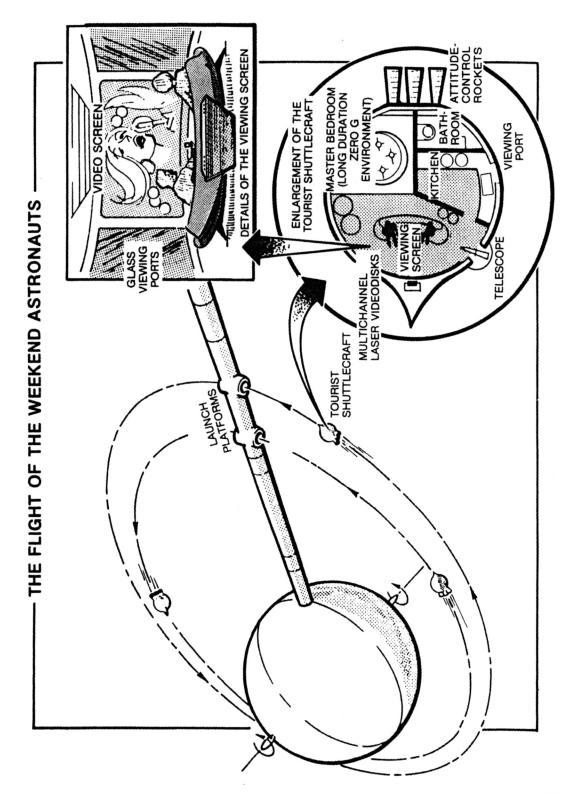

THE FLIGHT OF THE WEEKEND ASTRONAUTS

VIDEO SCREEN

GLASS VIEWING PORTS

DETAILS OF THE VIEWING SCREEN

ENLARGEMENT OF THE TOURIST SHUTTLECRAFT

MASTER BEDROOM (LONG DURATION ZERO G ENVIRONMENT)

ATTITUDE-CONTROL ROCKETS

BATH-ROOM

KITCHEN

VIEWING PORT

VIEWING SCREEN

TELESCOPE

MULTICHANNEL LASER VIDEODISKS

TOURIST SHUTTLECRAFT

LAUNCH PLATFORMS

circles inside a gas-filled dome. Here on Earth, we are unable to fly under our own power because our bodies are too heavy in relation to our strength. But, at the one-tenth g-levels provided by the Skyhook Complex, we could easily achieve self-propelled flight.

At the 32,000-mile altitude, a waste-disposal catipult will allow tomorrow's sanitation engineers to release radioactive wastes and other dangerous industrial by-products into safe orbits around the Sun. Launch platforms will be installed nearby for the release of payloads bound for the Moon, the planets, and the asteroid belt sandwiched between Jupiter and Mars.

Small, unpowered shuttlecrafts will be released from the 16,600-mile level to carry adventuresome tourists on one-day and two-day free-fall vacations around Earth. Figure 10.11 highlights some of the interesting features of a typical shuttlecraft. Notice how it is rigged with multichannel, wide-screen video disks synchronized with the current location of the craft.

The tourists on board will sit on special couches where they will see close-up versions of any tourist attractions they are flying over. The shuttlecraft will travel around Earth along a free-fall weightless trajectory and it will carry ample supplies of vintage wines and gourmet foods. When it becomes fully operational, young and old alike will arrive in a steady stream to experience the Skyhook Complex. Eventually, as its population begins to approach 100,000, its leaders may even campaign to become the fifty-first state.

Over the past three decades, my career has been structured in accordance with the dictum of Bertrand Russell: "When you are young and vigorous you do mathematics," he once wrote. "In middle age, you do philosophy. And, in your dotage, you write novels." My dotage years, it seems, have now officially arrived. In late fall of 1996, I began writing my first novel. It is a futuristic science fiction murder mystery set in outer space laced with tender romance and exciting chase scenes. It is entitled *Homicide at Skybook 1*. If you don't read the book, at least go to see the movie!

BIBLIOGRAPHY

CHAPTER 1: AN INTRODUCTION TO ORBITAL MECHANICS

Baker, M. L., Jr., and Maud W. Makemson. *An Introduction to Astrodynamics.* Academic Press, New York, 1960.

Bate, Roger R., Donald D. Mueller, and Jerry E. White. *Fundamentals of Astrodynamics.* Dover, New York, 1971.

Battin, Richard. *An Introduction to the Mathematics and Methods of Astrodynamics.* American Institute of Aeronautics and Astronautics, New York, 1987.

Bryan, C. D. B. *The National Air and Space Museum.* Abrams, New York, 1979.

Chobotov, Vladimir A. *Orbital Mechanics.* American Institute of Aeronautics and Astronautics, Washington, D.C., 1991.

Damon, Thomas D. *Introduction to Space: The Science of Spaceflight.* Orbit, Malabar, Fla., 1989.

Danby, J. M. A. *Fundamentals of Celestial Mechanics.* Macmillan, New York, 1964.

Epstein, Lewis Carroll. *Thinking Physics is Gedanken Physics.* Insight Press, San Francisco, 1990.

Flaste, Richard. *The New York Times Book of Science Literacy.* Times Books, New York, 1991.

Fortescue, Peter W., and John P. W. Stark. *Spacecraft Systems Engineering*, John Wiley, New York, 1991.

Hawkins, Stephen. *A Brief History of Time.* Bantam, New York, 1988.

Larson, Wiley J., and James R. Wertz. *Space Mission Analysis and Design.* Kluwer, Dordrecht, 1991.

Logsdon, Thomas S. *Breaking Through.* Addison-Wesley, Reading, Mass., 1993.

_____. *The Rush Toward the Stars.* Franklin, Englewood, N.J., 1970.

_____. *Space Industrialization: Executive Summary*, Final Report SP-78-AP-0055-1. Rockwell International Space Division, Downey, California, 1978, pp. 1–39.

Madewell, J. F., and Thomas S. Logsdon. "Space Flight Opportunities for Industry," paper read at the Space Flight Congress, Cape Kennedy, Fla., April 12, 1977.

Madonna, Richard. *Orbital Mechanics*. Krieger, Melbourne, Fla., 1991.

March, Robert H. *Physics for Poets*. Contemporary Books, Chicago, 1978.

Prussing, John E., and Bruce A. Conway. *Orbital Mechanics*. Oxford University Press, New York, 1993.

Sagan, Carl. *Cosmos*. Random House, New York, 1980.

Tufte, Edward R. *Envisioning Information*. Graphics Press, Cheshire, Conn., 1990.

_____. *The Visual Display of Quantitive Information*. Graphics Press, Cheshire, Conn., 1983.

CHAPTER 2: SATELLITE ORBITS

Baker, M. L., Jr., and Maud W. Makemson. *An Introduction to Astrodynamics*. Academic Press, New York, 1960.

Bate, Roger R., Donald D. Mueller, and Jerry E. White. *Fundamentals of Astrodynamics*. Dover, Inc., New York, 1971.

Battin, Richard. *An Introduction to the Mathematics and Methods of Astrodynamics*. American Institute of Aeronautics and Astronautics, New York, 1987.

Bryan, C. D. B. *The National Air and Space Museum*. Abrams, New York, 1979.

Chobotov, Vladimir A. *Orbital Mechanics*. American Institute of Aeronautics and Astronautics, Washington, D.C., 1991.

Curtis, Anthony R. *Space Almanac*. Gulf, Houston, 1990.

Feynman, Richard P., Robert B. Leighton, and Matthew Sands. *Lectures on Physics*. Addison-Wesley, Reading, Mass., 1963.

Johnson, Nicholas L., and David M. Rodvold. *Europe and Asia in Space*. Kaman Sciences Corp., Colorado Springs, 1991–1992.

Larson, Wiley J., and James R. Wertz. *Reducing Space Mission Cost*. Microcosm, Torrance, Calif., and Kluwer, Dordrecht, 1996.

Logsdon, Thomas S. "Opportunities in Space Industrialization," *Journal of Contemporary Business*. Vol. 7, No. 3, 1984, pp. 171–184.

_____. *The Rush Toward the Stars*. Franklin, Englewood, N.J., 1970.

Newton, Isaac. *Principia.* University of California Press, Berkeley and Los Angeles, 1934 (orginally published in 1686).

Prussing, John E., and Bruce A. Conway. *Orbital Mechanics.* Oxford University Press, New York, 1993.

Roddy, Dennis. *Satellite Communications.* Prentice Hall, Englewood Cliffs, N.J., 1989.

Wells, David, et al. *Guide to GPS Positioning.* Canadian GPS Associates, Fredericton, New Brunswick, 1987.

Yenne, Bill. *The Encyclopedia of US Spacecraft.* Exeter, New York, 1985.

CHAPTER 3: THE ORBITAL ENVIRONMENT

Damon, Thomas D. *Introduction to Space: The Science of Spaceflight.* Orbit, Malabar, Fla., 1989.

Eckart, Peter. *Spaceflight Life Support and Biospherics.* Microcosm, Torrance, Calif., and Kluwer, Dordrecht, 1996.

Fortescue, Peter W., and John P. W. Stark. *Spacecraft Systems Engineering.* John Wiley, New York, 1991.

Friedman, Herbert. *Sun and Earth.* Scientific American, New York, 1986.

Heppenheimer, T. A. *Toward Distant Suns.* Stackpole, Harrisburg, Penn., 1979.

Johnson, Nicholas L., and Darren S. McNight. *Artificial Space Debris.* Orbit, Malabar, Fla., 1987.

Larson, Wiley J., and James R. Wertz. *Space Mission Analysis and Design.* Kluwer, Dordrecht, 1991.

Logsdon, Thomas S. *Mobile Communication Satellites: Theory and Applications.* McGraw-Hill, New York, 1995.

————. "The Practical Benefits of Hydrogen Masers in Space," paper read at the Second International Conference on Frequency Control and Synthesis, University of Leicester, England, April 10–13, 1989.

Maurer, Richard. *Junk in Space.* Simon & Schuster, New York, 1989.

"Nuclear-Powered Spacecraft Deployed into Earth Orbit." *Scientific American,* June 1991, p. 46.

Ordway, Frederick I., III, and Ernst Stuhlinger. *Wernher von Braun: Crusader for Space.* Krieger, Malabar, Fla., 1994.

Sagan, Carl. *Pale Blue Dot, A Vision of the Human Future in Space.* Random House, New York, 1994.

Sarafin, Thomas P., and Wiley J. Larson. *Spacecraft Structures and Mechanisms—From Concept to Launch.* Microsm, Torrance, Cal., and Kluwer, Dordrecht, 1995.

Shipman, Harry L. *Space 2000, Meeting the Challenge of a New Era.* Plenum Press, New York, 1987.

Tribble, Alan C. *The Space Environment, Implications for Spacecraft Design.* Princeton University Press, Princeton, 1995.

Wertz, James R. *Spacecraft Attitude Determination and Control.* Kluwer, Dordrecht, 1995.

CHAPTER 4: POWERED FLIGHT MANEUVERS

Baker, M. L., Jr., and Maud W. Makemson. *An Introduction to Astrodynamics.* Academic Press, New York, 1960.

Bate, Roger R., Donald D. Mueller, and Jerry E. White. *Fundamentals of Astrodynamics.* Dover, New York, 1971.

Battin, Richard. *An Introduction to the Mathematics and Methods of Astrodynamics.* American Institute of Aeronautics and Astronautics, New York, 1987.

Chobotov, Vladimir A. *Orbital Mechanics.* American Institute of Aeronautics and Astronautics, Washington, D.C., 1991.

Cohen, Barbara. *Pioneering the Space Frontier, The Report of the National Commission on Space.* Bantam, New York, 1986.

Fortescue, Peter W., and John P. W. Stark. *Spacecraft Systems Engineering.* John Wiley, New York, 1991.

Larson, Wiley J. and James R. Wertz. *Space Mission Analysis and Design.* Kluwer, Dordrecht, 1991.

Ley, Willy. *Missiles, Moon Probes, and Megaparsecs.* Signet, New York, 1962.

Logsdon, Thomas S. *The Rush Toward the Stars.* Franklin, Englewood, N.J., 1970.

Pattan, Bruno. *Satellite Systems: Principles and Technologies.* Van Nostrand Reinhold, New York, 1993.

Pocha, J. J. *An Introduction to Mission Design for Geostationary Satellites.* D. Reidel, Dordrecht, 1987.

Prussing, John E., and Bruce A. Conway. *Orbital Mechanics.* Oxford University Press, New York, 1993.

Sagan, Carl. *Pale Blue Dot, A Vision of the Human Future in Space.* Random House, New York,1994.

Soop, E. M. *Handbook of Geostationary Orbits.* Kluwer, Dordrecht, and Microcosm, Torrance, Calif., 1994.

Thompson, William Tyrell. *Introduction to Space Dynamics*. Dover, New York, 1990.

CHAPTER 5: BOOSTING A SATELLITE INTO ORBIT

Damon, Thomas D. *Introduction to Space: The Science of Spaceflight*. Orbit, Malabar, Fla., 1989.

Fortescue, Peter W., and John P. W. Stark. *Spacecraft Systems Engineering*. John Wiley, New York, 1991.

Hunter II, Maxwell W. *Thrust Into Space*. Holt, Rinehart and Winston, New York, 1966.

Isakowitz, Steven J. *International Reference Guide to Space Launch Systems*. American Institute of Aeronautics and Astronautics, New York, 1991.

Logsdon, Thomas. *The Rush Toward the Stars*. Franklin, Englewood, N.J., 1970.

Pattan, Bruno. *Satellite Systems: Principles and Technologies*. Van Nostrand Reinhold, New York, 1993.

Sarafin, Thomas P., and Wiley J. Larson. *Spacecraft Structures and Mechanisms—From Concept to Launch*. Microsm, Torrance, Calif., and Kluwer, Dordrecht, 1995.

Sutton, George P. *Rocket Propulsion Elements: An Introduction to the Engineering of Rockets*. John Wiley, New York, 1986.

CHAPTER 6: TODAY'S FAMILY OF GLOBAL BOOSTERS

Asker, James R. "Official Urges U.S. to License Ariane 5." *Aviation Week and Space Technology*, April 10, 1995.

————. "Racing to Make Space Economics." *Aviation Week and Space Technology*, April 3, 1995.

Bryan, C. D. B. *The National Air and Space Museum*. Abrams, New York, 1979.

Damon, Thomas D. *Introduction to Space: The Science of Spaceflight*. Orbit, Malabar, Fla., 1989.

De Selding, Peter. "Five Launchers Competing for Full-Sized Payloads," *Space News*, May 8-14, 1995.

Edelson, Burton, et al. *Satellite Communications Systems and Technology, Vol. 1: Analytical Chapters*. International Technology Research Institute, Baltimore, 1993.

Ferster, Warren. "Delta-Lite Launcher to Compete with LLV," *Space News*, June 19-25, 1995.

_____. "Industry Steps Up to Reusable Rocket Efforts," *Space News*, March 13-19, 1995.

Fortescue, Peter W., and John P. W. Stark. *Spacecraft Systems Engineering*. John Wiley, New York, 1991.

Iannotta, Ben. "U.S. to Revamp Resuable, Expendable Fleets," *Space News*, May 8-14, 1995.

Isakowitz, Steven J. *International Reference Guide to Space Launch Systems*. American Institute of Aeronautics and Astronautics, New York, 1991.

_____. *International Reference Guide to Space Launch Systems*. American Institute of Aeronautics and Astronautics, Washington, D.C., 1995.

Keith, Edward L. "Low-Cost Space Transportation: The Search for the Lowest Cost." *American Astronautical Society/American Institute of Aeronautics and Astronautics*. Technical paper, February 13, 1991, pp. 1-20.

Kerrod, Robin. *Space Shuttle*. Gallery, New York, 1984.

Klevatt, Gaubatt. "Single Stage Rocket Technology," paper read at the 43rd Congress of the International Federation, Washington, D.C., August 28-30, 1995.

Popelewski, Robert. "Reusable Launcher Concepts Pursued," *Interavia*, May, 1995.

CHAPTER 7: ENHANCING THE PERFORMANCE OF BOOSTER ROCKETS

Africano, R. C., and T. S. Logsdon. *Enhancing the Saturn V Moon Rocket's Translunar Payload Capability*, Technical Paper No. 69-451. American Institute of Aeronautics, Colorado Springs, 1969.

Indrikis, Janis, and Jeffrey C. Preble. "Bridging the Gap Between High and Low Acceleration for Planetary Escape," paper read at the American Astronautical Society/American Institute of Aeronautics Spaceflight Meeting, Colorado Springs, Feb. 24-26, 1992.

Larson, Wiley J., and James R. Wertz. *Reducing Space Mission Costs*. Microcosm, Torrance, Calif., and Kluwer, Dordrecht, 1996.

Logsdon, Thomas. *Breaking Through*. Addison-Wesley, Reading, Mass., 1993.

Ordway, Frederick I., III and Ernst Stuhlinger. *Wernher von Braun: Crusader for Space*. Krieger, Malabar, Fla., 1994.

"Russian Liquid Rocket Bureau Designing Three-Fuel Engine," *Aviation Week and Space Technology*, March 30, 1992, p. 24.

Thompson, W. T. *Introduction to Space Dynamics*. John Wiley, New York, 1961.

CHAPTER 8: CHOOSING THE PROPER ORBIT FOR A SATELLITE

Alper, Joel, and Joseph N. Pelton. *The Intelsat Global Satellite System*. American Institute of Aeronautics and Astronautics, New York, 1984.

Cohen, Barbara. *Pioneering The Space Frontier, The Report of the National Commission on Space*. Bantam, New York, 1986.

Damon, Thomas D. *Introduction to Space: The Science of Spaceflight*. Orbit, Malabar, Fla., 1989.

Drain, J. *Lightsat Constellation Designs*. American Institute of Aeronautics and Astronautics, New York, 1992.

Drain, John E., and Thomas J. Kacena. *Populating the Abyss-Investigating More Efficient Orbits: Getting More Miles per Gallon for Your (Space) Vehicle*, prepublication booklet. Space Applications Corp., 1992, pp. 1-15.

Edelson, Burton, et al. *Satellite Communications Systems and Technology, Vol. 1: Analytical Chapters*. International Technology Research Institute, Baltimore, 1993.

Einstein, Albert. *The Meaning of Relativity*, 5th ed. Princeton University Press, Princeton, 1956.

Farquahr, Robert W. "Future Missions for Libration Point Satellites," *Astronautics and Aeronautics*. May 1969, pp. 52-55.

_____. "A Halo Orbit Lunar Station," *Astronautics and Aeronautics*, June 1972, pp. 59-63.

Fitz-Randolph, Jane, and James Jespersen. *From Sundials to Atomic Clocks*. U.S. Department of Commerce, National Bureau of Standards, Washington, D.C., 1977.

Fritzsch, Harald. $E=MC^2$: *An Equation That Changed the World*. University of Chicago Press, Chicago, 1994.

Jorgensen, Paul S. "Special Relativity and Intersatellite Tracking," *Navigation: Journal of the Institute of Navigation*, Winter 1988-1989, pp. 429-442.

Kaplan, Elliot D. *Understanding GPS Principles and Applications*. Artech House, Norwood, Mass., 1996.

Larson, Wiley J., and James R. Wertz. *Space Mission Analysis and Design*. Kluwer, Dordrecht, 1991.

Logsdon, Thomas S., and C. W. Helms. "The Navstar GPS: A Status Report," paper read at the 5th Annual Armed Forces Communications and Electronics Association Symposium and Exposition, October 24, 1984, Brussels.

Logsdon, Thomas S. *Mobile Communication Satellites: Theory and Applications*. McGraw-Hill, New York, 1995.

_____. *The Navstar Global Positioning System*. Van Nostrand Reinhold, New York, 1992.

_____. "Orbiting Switchboards." *Technology Illustrated*, October-November 1981, pp. 54-62.

_____. *The Rush Toward the Stars*. Franklin, Englewood, New Jersey, 1970.

_____. "Unsnarling Signals in Space," *Technology Review*, August-September 1982, pp.14-18.

Macaulay, David. *The Way Things Work*. Houghton Mifflin, Boston, Mass.,1988.

McGuinness, Michael, and Joseph Schwartz. *Einstein for Beginners*. Pantheon, New York, 1979.

"New Timetables for Planetary Tours," *Time Magazine*, August 23, 1968, p. 39.

Pattan, Bruno. *Satellite Systems: Principles and Technologies*. Van Nostrand Reinhold, New York, 1993.

Pocha, J. J. An Introduction to Mission Design for Geostationary Satellites. D. Reidel, Dordrecht, 1987.

Roddy, Dennis. *Satellite Communications*. Prentice Hall, Englewood Cliffs, N.J., 1989.

Soop, E. M. *Handbook of Geostationary Orbits*. Kluwer, Dordrecht, and Microcosm, Torrance, Calif., 1994.

Turner, Andrew E., and Kent M. Price. "The Potential for Non-Geosynchronous Orbit," *Satellite Communication*, June, 1988.

Wells, David, et al. *Guide to GPS Positioning*. Canadian GPS Associates, Fredericton, New Brunswick, 1987.

Wertz, James R. *Spacecraft Attitude Determination and Control*. Kluwer, Dordrecht, 1995.

CHAPTER 9: CHOOSING THE PROPER CONSTELLATION ARCHITECTURE

Alper, Joel, and Joseph N. Pelton. *The Intelsat Global Satellite System*. American Institute of Aeronautics and Astronautics, New York, 1984.

Anodina, T. G. "Global Positioning System Glonass," paper read at the 4th Meeting of the Special Committee on Future Air Navigation Systems (FANS), Montreal, May 2-20, 1988.

Indrikis, Jamis, and Robert Cleave. "Space Eggs," paper read at the 5th Annual Utah State University Conference on Small Satellites, Logan, Utah, August 28, 1991, pp. 1-7.

Johnson, Nicholas L., and David M. Rodvold. *Europe and Asia in Space*. Kaman Sciences Corp., Colorado Springs, 1991-1992.

Larson, Wiley J., and James R. Wertz. *Space Mission Analysis and Design*. Kluwer, Dordrecht, 1991.

Leick, Alfred. *GPS Satellite Surveying*. John Wiley, New York, 1995.

Logsdon, Thomas S. *Mobile Communication Satellites: Theory and Applications*. McGraw-Hill, New York, 1995.

———. *The Navstar Global Positioning System*. Van Nostrand Reinhold, New York, 1992.

———. "Satellites Bring New Precision to Navigation," *High Technology*, July-August 1984, pp. 61-66.

Logsdon, Thomas S., and Charles W. Helmes. "Promising Third-World Applications of the Navstar Global Positioning System," paper read at the Institute of Navigation Fortieth Annual Meeting, Cambridge, Mass., June 1984.

Misva, J. G., et al. "Glonass Data Analysis: Interium Results." *Navigation: Journal of the Institute of Navigation*, Spring 1992, pp. 93-109.

Pattan, Bruno. *Satellite Systems: Principles and Technologies*. Van Nostrand Reinhold, New York, 1993.

Roddy, Dennis. *Satellite Communications*. Prentice Hall, Englewood Cliffs, N.J., 1989.

Rusch, Roger J. *Comparison of Personal Communication Satellite Systems*. TRW, Redondo Beach, Calif., 1993.

Soop, E. M. *Handbook of Geostationary Orbits*. Kluwer, Dordrecht, and Microcosm, Torrance, Calif., 1994.

Walker, J. G. *Continuous Whole-Earth Coverage by Circular-Orbit Satellite Patterns*, Technical Report 77044. Royal Aircraft Establishment, Farnborough, England, 1977.

Wells, David, et al. *Guide to GPS Positioning*. Canadian GPS Associates, Fredericton, New Brunswick, 1987.

Wertz, James R. *Spacecraft Attitude Determination and Control*. Kluwer, Dordrecht, 1995.

Williams, J. E. D. *From Sails to Satellites*. Oxford University Press, New York, 1992.

CHAPTER 10: SPACE-AGE TECHNOLOGIES FOR THE TWENTY-FIRST CENTURY

"Advanced Propulsion on a Shoestring," *Aerospace America*, May, 1990.

Asker, James R. "Atlantis to Evaluate Characteristics of Tethered Satellite." *Aviation Week and Space Technology*, June 20, 1992, pp. 40-44.

Augustine, Norman R. *Augustine's Laws and Major System Development Programs*. American Institute of Aeronautics and Astronautics, New York, 1983.

Bryan, C. D. B. *The National Air and Space Museum*. Abrams, New York, 1979.

Curtis, Anthony R. *Space Almanac*. Gulf, Houston, 1990.

Hawkins, Stephen. *A Brief History of Time*. Bantam, New York, 1988.

Heppenheimer, T. A. *Colonies in Space*. Stackpole, Harrisburg, Penn., 1977.

_____. *Toward Distant Suns*. Stackpole, Harrisburg, Penn., 1979.

Isaacs, John D., et al., "Satellite Elongation Into a True Skyhook," *Science Magazine*, Vol. 151, February 11, 1966, pp. 682-683.

Logsdon, Thomas S. "High Fliers," *Technology Illustrated*, December-January 1982, pp. 24-32.

_____. *Industries in Space to Benefit Mankind*, Report SP77-AP-0094. Rockwell International Space Divisioln, 1977.

_____. *Mobile Communication Satellites: Theory and Applications*. McGraw-Hill, New York, 1995.

_____. *The Rush Toward the Stars*. Franklin, Englewood, N.J., 1970.

Lvov, Vladimir. "Sky-Hook Old Idea," *Science Magazine*, Vol. 158, November 17, 1969, pp. 946-947.

Petroski, Henry. *Beyond Engineering*. St. Martin's Press, New York, 1977.

Sagan, Carl. *Cosmos*. Random House, New York, 1980.

Shipman, Harry L. *Space 2000, Meeting the Challenge of a New Era*. Plenum Press, New York, 1987.

"Skyhook' Satellite Tied to Earth Suggested as Launching Platform." Associated Press, February 2, 1966.

Sutton, G. W. "Synchronous Rotation of a Satellite at Less Than Synchronous Altitude," *AIAA Journal*, Vol. 5, No. 4, April 1967, pp. 813-815.

ABOUT THE AUTHOR

For 35 years, Tom Logsdon worked as an orbital mechanics specialist at McDonnell Douglas and at Rockwell International. His most stimulating assignments included six-degree-of-freedom trajectory simulations and powered flight maneuver sequences for Project Apollo and rendezvous studies for the manned Skylab. He also carried out various constellation selection trades and systems analysis studies for the twenty-four satellite Navstar Global Positioning System (GPS).

During those challenging years, Logsdon wrote dozens of technical reports and delivered more than 600 engineering presentations to appreciative audiences ranging from Boy Scout troops to the National Science Foundation.

His honors have included Rockwell's Sustained Superior Performance Award for Engineering Excellence, their cash Ideas and Inventions Award, and the company's prestigious Presidential Award. In 1984, he was chosen as Alumni of the Year at Eastern Kentucky University from among 45,000 eligible candidates. He was also rated as one of the "Top Ten" professional platform lecturers at the International Platform Association's 158th annual meeting in Washington, D.C.

Logsdon has written forty technical papers and journal articles and twenty-nine nonfiction books. One of them, *The Robot Revolution* was a Book-of-the-Month Club selection. Another, his playful trade book *How to Cope with Computers*, was chosen by *Publisher's Weekly* as one of 100 best technical books of 1982. And his textbook *The Microcomputer Explosion* received a special award from the National Press Club for its exceptional graphics.

Logsdon's professional platform lectures and his three-day short courses on "Orbital Mechanics," "Mobile Communication Satellites," "Team-Based Problem Solving," and "The Navstar GPS" have been attended by engineers, managers, and scientists in eighteen different countries scattered across five continents. When he is not writing and revising technical books on commercial jetliners at 36,000 feet, Logsdon lives on the Pacific coast in Seal Beach, California. There he thrives on Hollywood movies, touring Broadway plays, and spastic tennis games. He also enjoys his perky wife Cyndy and his lively stepson Chad.

INDEX

Absent-minded professor, 29
Ace-orbit constellations, 185
Action-reaction, 17
Adding lightness, 121–124
Africano, Bob, 242–243, 245
Aldrin, Edwin, 149, 168
America's expendable boosters, 132–136
Apogee, 43
Apollo, 13, 27
Apollo capsules, 76
Apsidal rotation 60–62
Area calculations, 217
Argument of perigee, 53–54
Ariane booster, 136–137
Ariane V, 140
Aristotle, 3, 8, 17
Armstrong, Neil, 149, 168
Ascending node, 54
Asparagus-stalk booster, 143–144
Astronomical unit, 4, 7
Atlas family of boosters, 133
Atlas II, 134
Atmospheric density, 67–68
Atmospheric drag, 58, 69
Atmospheric refraction, 9

Ballistic capture missions, 103–105
Ballistic parameter, 67
Balloon-tank design, 122
Barycenter, 18
Battin, Richard, H., 52
Belbruno and Miller, 103
Beneficial properties of space, 64
Bernoulli, John, 158, 160
Bernoulli brothers, 158, 160
Beta angle, 172
Biasing to maximize payload, 158–159
Biasing to minimize residuals, 158

Bielliptic transfer maneuver, 90–92
Big, dumb booster, 142–143
Biparabolic transfer, 92
Booster rockets, 127
Brachistochrone problem, 159–163
Brahe, Tycho, 8, 9, 19
Brux, Bud, 153–154, 156
Burning rates, 120
Burning-mixture ratio, 153

Calculating the orbit of Mars, 10
Capacitance probe, 153
Centrifugal force, 20
Ceres, 165
Chinese fire arrows, 112
Chinese Long March, 139
Cigarette-end burning, 112
Circular orbit, 30–31, 76
Circular orbital velocity, 35
Cleeve, Bob, 218
Collinear libration points, 191, 200
Collision energy for space debris, 77–78
Columbus, Christopher, 238
Combined plane change maneuvers, 88
Comet trails, 74–76
Common bulkhead, 122–124
Communication satellites, 65–66, 185
Computer-modeling techniques, 215–221
Conic sections, 18, 23
Constellation architecture, 207–209
Constellation selection trades, 220–230
Construction of the Skyhook pipeline, 248
Copernicus, Nicolaus, 3–5, 7, 9, 19
Coplanar Hohmann transfer maneuvers, 82–83
Cost of accelerating propellants to high velocities, 41
Cost of accelerating a rocket's unburned propellants, 124–126

Cost of a constellation, 223
Cost of launching a satellite, 40
Coverage analysis, 217–221
Coverage footprints, 218, 220
Crosslinks, 225
Cycloid-shaped sliding board, 162

Dawn-dusk orbit, 176
Dawn-dusk sun-synchronous orbit, 173
Debris fragments, 77
Decay-rate history for a satellite, 67–68
Delta II Medium Launch Vehicle, 132
Delta III booster, 140–141
Deorbit maneuvers, 97
Deorbit velocity increment, 100
Deorbiting the S-II stage, 97, 101
Direct power law of gravitation, 14
Doppler-shift variations, 169–172
Drag coefficient, 67
Drag force acting on a satellite, 67
Drag losses, 119, 155
Draim, John, 211, 214–215, 217
Drop towers, 26–27, 64
Dropping satellites in orbit, 246
Due east launch, 56–57
Dula, Art, 231
Dyson, Freeman, 234–235

Earth Radiation Budget Satellite, 189
Earth's atmosphere, 67, 70
Earth's equatorial bulge, 58–59
Earth's gravitational field, 69
Earth-seeing orientation, 71
Eccentric anomaly, 46–52
Eccentricity, 53–54
Echo balloon, 65
Eclipse intervals, 223–224
Ecliptic plane, 4
Electromagnetic catapults, 233–234
Electromagnets, 74
Elevation angles, 223
Elliptical orbits, 10, 22, 29–30, 37, 43, 53, 59
Energy costs, 40
Energy generation using the skyhook, 247–249
Ephermeris constants, 53
Epicycles, 2, 4
Equal-period descent orbit, 199–201
Equatorial bulge, 60, 70
Equilateral libration points, 191, 201
Escape trajectory, 30–32, 45
Escape velocity, 35
Escaping from an asteroid, 34, 36

European Ariane, 79, 136–137
Exhaust molecules, 108, 110
Exhaust nozzle, 107
Explosions, 76–80
Exterior planets, 5

Falling moon, the, 1, 12–13
Farquahr, Robert, 192
Fiction Skyhook book, 252
Figure-8 ground trace, 178
Flight of the weekend astronauts, 251
Flight-path angle, 101
Forward, Robert, 240
Free-return figure 8 trajectory, 199–201
Frozen orbits, 189
Fuel biasing, 157
Full-sun, sun-synchronous orbits, 175
Fundamentals of space exploration, 242–243

Galilei, Galileo, 7–8, 13, 19, 21
Gauss, Karl Frederich, 163, 165
Generalizing Kepler's laws, 17–18
Geosynchronous altitude, 45
Geosynchronous satellites, 78, 93, 95, 175–183, 196
Glenn, John, 133
Global boosters, 127
Globalstar mobile communication satellites, 135
Glonass satellites, 209
Goddard, Robert, 109–111
Goode equal-area projection, 216
Grand tour missions, 195–196
Gravitational acceleration, 14–15
Gravitational field, 69
Gravitational parameter, 32–34, 36
Gravitational perturbations, 58
Gravity inside the earth, 14
Gravity losses, 84, 119, 155
Gravity wells, 44, 46–52, 103, 105
Gravity-gradient stabilization, 71–72, 171–172
Grid search, 217
Ground-trace geometry, 56–58
Growth rates for space debris, 77

Halley, Edmund, 29, 81
Halley's comet, 29
Halo orbits, 192
Handoffs, 225
Hard vacuum, 64–65
Hardin, Johnny, 63, 66
Heliocentric theory of Copernicus, the, 3–5
Higher-density propellants, 124
Higher-order shape terms, 70

Hill-Clohessy-Wiltshire relative-motion equations, 95
Hiten spacecraft, 104
Hohmann, Walter, 81–82, 192
Hohmann transfer maneuver, 81–82, 88, 90, 103, 193
Homicide at Skybook I, 252
Houbolt, John, 152
Hyper-velocity impacts, 78–79

IBM 7094 mainframe computer, 166
Ideal velocity, 155
Inclination, 53–54
Inclined geosynchronous orbits, 175–177
Incremental startup, 225
Indian rope trick, 243–244
Indrikis, Janis, 218
Industrial Revolution, 8
Inflatable structures, 65
Intelsat communication satellites, 205, 208
Intelsat constellation, 207
Interior phasing orbit, 96
Interior planets, 4
Internal and external phasing orbits, 94
Interplanetary trajectories, 192–195
Inverse square gravitational law, 12–15
Iridium constellation, 172
Isaacs, John D., 244–245
Iterative least squares hunting procedure, 166, 168

Japanese H-2, 139
Jefferson, Thomas, 231

Keith, Ed, 143, 144, 146
Kennedy, President John F., 149
Kepler, Johannes, 10, 17–19, 45, 49
    Keplerian orbital elements, 53–55, 58
    Kepler's equation, 45–53
    Kepler's law of equal areas, 22, 37, 51
    Kepler's three laws of planetary motion, 9–11, 17–18
Keynes, John Maynard, 201
Killer satellites, 77
Kinetic energy, 38, 42–43

Lagrange, Joseph-Louis, 190–191, 200
Landsat constellation, 205, 207
Landsat Earth resources satellites, 189, 208
Langley, Samuel, 128
Lapping interval, 7
Laser powered elevators, 247

Laser-powered rockets, 236–237
Launch azimuth, 56–58
Launch boundaries, 57–58
Launching a satellite into orbit, 24
Law of equal areas, 10
Law of uniform acceleration, 8, 13
Laws of motion, 16
Leaning Tower of Pisa, 7
Leaping lizard, 161–163
Ley, Willy, 91
Libration-point orbits, 190–192, 199–201
Liquid-fueled rockets, 110–111
Lodestones, 71
Long March booster, 139
Low-thrust spiraling trajectories, 163
Lunar-orbit rendezvous, 150, 152
Lunar-solar perturbations, 177–182

Maglifter, 144–146
Magnetic field, 71
Magnetic momentum dumping, 72
Magnetically levitated booster, 146
Man-made space debris, 76
Map projections, 216
Mars-mission opportunities, 193
Mass drivers, 231
Mass fraction, 116
McClure, Frank, 169
Mean anomaly, 46–52
Mercator projection, 216
Mercury astronauts, 24
Meteor Crater, 74
Meteor showers, 74
Meteoroids, 74–75
Mickey Mouse balloon filled with pink lemonade, 122
Microgravity, 27, 64
Micrometeorites, 76
Mobile communication satellites, 220–230
Mollweide equal-area projection, 216
Molniya communication satellites, 62, 185–188, 198, 201, 207
Molniya orbits, 185
Momentum exchange device, 108–109
Momentum wheels, 73
Monte Carlo simulation, 156–158
Mountain climbing expeditions, 116
Multi-impulse low-thrust maneuvers, 84–85
Multistage booster rockets, 41, 116–118, 124
Munitions makers, 64
Mystery of the wandering stars, 2

Navstar Global Positioning System (GPS), 58, 60, 92, 96, 132-133, 184, 204-205
Nelson, Ed, 45
Nerva rockets, 234
Newman, Edwin, 130
Newton, Isaac, 1-2, 8, 11, 13, 17, 19, 20-21, 24-25, 29-30, 37, 42, 51, 53, 63, 69-70, 81, 107-108, 115, 127, 149, 158-163, 169, 203, 231
  and the apple, 12
  childhood, 12
  laws of motion, 16-17
  method, 51
  pitch to orbit, 25
  second law, 17
  universal law of gravitation, 12-15
Nodal regression, 59
Nuclear propulsion, 234
Nullifying the effects of orbital perturbations, 182

Oblate spheroid, 70
O'Neill, Gerard, 233
On-orbit rendezvous, 94
Optimal fuel biasing, 156-157, 166-167
Optimal trajectory shaping, 158-163
Orbital decay rates, 68
Orbital eccentricity, 54
Orbital elements, 53-55, 58
Orbital environment, 63, 222
Orbital inclination, 54
Orbital perturbations, 58-62
Orbital velocity, 30
Orthographic projection, 216
Outer planets, 7
Oxidizer residuals, 156

Parabolic escape trajectory, 30-32
Pegasus booster, 134-136
Pendulum principle, 8
Perigee, 43, 46
Personal communicators, 225, 227
Perturbations acting on a geosynchronous satellite, 177, 183
Phasing orbit, 94
Plane-change maneuvers, 85-90
Planetary swing-by maneuvers, 101-102
Plumber's nightmare, 111
Polar birdcage constellations, 169-172
Polar orbits, 169-172
Polesats, 238-240
Pope, Alexander, 107
Position and velocity of a satellite, 55

Postflight trajectory reconstruction, 163-168
Potential energy, 38, 42-43
Potential harmonics, 58, 70
Powered flight maneuvers, 44
Preflight trajectory simulations, 164
Principia, 16
Programmed Mixture Ratio Scheme, 153-156, 167-168
Project Apollo, 150, 242
Project Orion, 234-235
Project Skyhook, 242, 246
Project West Ford, 209-210
Propellant Utilization System, 151, 153
Ptolemy's earth-centered system, 2-3

Radioactive payloads, 79-80
Reaction wheels, 73
Regatta satellites, 237
Reilly, Dennis, 236-237
Relative motion coordinates, 95, 98-99
Relative-motion equations, 98-99
Rendezvous and docking studies, 95
Right ascension of the ascending node, 53-54
Rocket, 107-109
Rosette Constellations, 211
Rotating coordinate system 99
Rush toward the Stars, The, 127, 159
Russian booster rocket design, 137-138
Russian Energiya, 138
Russian space shuttle, 130-132

Satellite costs, 225
Satellite lifetime, 225
Satellite orbits, 29
Saturn V Moon rocket, 97, 123, 127, 134, 151, 154, 165-167
Scale height, 69
Schnitt, Arthur, 146
Semimajor axis, 53-54
Shooting stars, 74
Shot towers, 26-27, 64
Shuttle ascent trajectory, 164
Signal time delays, 223-224
Single-stage rockets, 116-118
Skyhook complex, 248, 250, 252
Skyhook pipeline, 246-247, 249
Skyhooks, 241
Skylab capsule, 97
Skylab insurance, 101
Solar radiation pressure, 58
Solar sails, 236-238
Solar-radiation pressure, 177-182, 239

Solid-fueled rockets, 111
Sounding rockets, 27, 45
South Atlantic Anomaly, 72–74
Soviet boosters, 137–138
Soviet Cicada, 205
Soviet Glonass, 205
Soviet Molniya, 206
Space Infrared Telescope Facility (SIRTF), 192
Space debris fragments, 76–80, 223–224
Space shuttle, 41, 71, 97, 128–130
Space-age slingshots, 101–103
Space-eggs computer-simulation program, 218–221
Spacecraft elevation angles, 224
Special orbits, 196–201
Specific energy, 37
Specific impulse, 113–114, 155
Splitting the plane change, 88–90
Sporadic meteoroids, 75
Sputnik, 1, 15–16, 63, 169
Staging techniques, 126
Star Trek, 22
Star-shaped perforation, 111
Stationkeeping velocity, 179
Steering angles, 119
Steering losses, 155
Stewart, Homer, 195
Strap-on solid-fuel rockets, 130
Sun-synchronous orbits, 172–175, 196
Sun-synchronous satellites, 174
Swing-by maneuvers, 75, 101–103
Syncom II, 177
Synoptic period, 7

Tapering the cable, 243–245
Taurus rocket, 135–136
Tesserial harmonics, 177–182
Tether ball analogy, 22
Tethered satellites, 240–242
Thermal tiles, 130
Three concepts for conquering the moon, 149
Time of perigee passage, 54
Titan booster, 134
Topex/Poseidon oceanographic satellite, 189

Trajectory losses, 118, 155
Trajectory shaping, 158–163
Transfer ellipse, 96
Transit navigation satellites, 70–71, 170–172,
    204–205
Translunar trajectory, 155
Transportation costs, 225
Tree-top satellites, 34–35
True anomaly, 38, 46–52
Tsiolkovsky's equation, 114–116
Two-stage booster rockets, 116–117

Van Allen radiation belts, 72–74, 222–224
Velocity assist from earth's rotational velocity, 57
Velocity increments needed to reach the various
    planets, 194
Velocity losses, 155
Velocity profiles, 120–122
Vernal equinox, 53
Verne, Jules, 232
Viewing contours, 220
Vis viva equation, 30–32, 34, 37–38, 40, 42
Von Braun, Werner, 149–150, 152
Von Leibniz, Gottfried Wilhelm, 1
Voyager space probe, 196–197

Walker, John, 211, 213
Walker constellations, 212–213
walking-orbit maneuvers, 92–96
Wandering stars, the, 2–4, 9, 160
Wayne, John, 243
Weak stability boundary, 104
Weight of the earth, 69
Weightlessness, 25–27
West Ford needles, 209–210
What is a constellation?, 204
What is a rocket?, 107
Why doesn't a satellite fall?, 20–21
Wide-angle view from space, 65
Wolfe, John, 158–159
Wordsworth, William, 149

X-33 reusable space plane, 139